STUDY SKILLS FOR GEOGRAPHY, EARTH AND ENVIRONMENTAL SCIENCE STUDENTS

Third Edition

Pauline E. Kneale

School of Geography,
Earth and Environmental Sciences
University of Plymouth, UK

HODDER
EDUCATION
AN HACHETTE UK COMPANY

First published in Great Britain in 1999

This third edition published in 2011 by
Hodder Education, an Hachette UK Company,
338 Euston Road, London NW1 3BH

Hachette UK's policy is to use papers that are natural, renewable and
recyclable products and made from wood grown in sustainable forests.
The logging and manufacturing processes are expected to conform to the
environmental regulations of the country of origin.

The advice and information in this book are believed to be true and
accurate at the date of going to press, but neither the author[s] nor the publisher
can accept any legal responsibility or liability for any errors or omissions.

British Library Cataloguing in Publication Data
A catalogue record for this book is available from the British Library

Library of Congress Cataloging-in-Publication Data
A catalog record for this book is available from the Library of Congress

ISBN: 978 1 444 120 967

1 2 3 4 5 6 7 8 9 10

Typeset in Garamond and Helvetica by Dorchester Typesetting, Dorchester, Dorset
Printed and bound in the UK by CPI Antony Rowe, Chippenham, Wiltshire

What do you think about this book? Or any other Hodder Education
Please send your comments to educationenquiries@hodder.co.uk

http://www.hoddereducation.com

Contents

Preface

Third Edition

Thinking of swimming in Paris?

This book is written for students studying Geography, Environmental Science, Earth Science (GEES) and related disciplines, for example Transport Studies, Tourism and related fieldwork-based disciplines: Ecology, Land Management and Landscape Studies. If you want to be a successful student but can't work out how university works, start here. The book aims to demystify some aspects of university life and study, and to build your confidence to learn effectively. The acronym GEES is used as shorthand for the three disciplines, 'GEES students' refers to geographers, environmental scientists and earth scientists. Students studying geography, earth or environmental science are likely to share classes with each other. There are a great many commonalities between the degrees both in terms of topic and the processes of study and research. There is not enough room for an example from each GEES discipline on every point: please remember that the concepts usually apply equally to all the GEES disciplines. Points made about environmental scientists are normally equally applicable to geologists and geographers.

You'll be declared in Seine.

Most people using skills and self-help books, or on training courses, find that 90 per cent of the material is familiar, but new ideas make it worthwhile. The 90 per cent increases your confidence that you are on the right track, and the remainder, hopefully, sparks some rethinking, reassessment and refining. You have been studying for 15 years at least, and are very skilled in various ways. Some approaches at university will be new; a quick glance here should help you to understand what's going on. The trick with university study is to find a combination of ideas that suit you, promote your research and learning, and enhance your self-confidence. As with all books, not all the answers are here, but who said study would be a doddle? This text is intended for reference throughout your degree; some items are irrelevant at first, but important later. Chapters are deliberately very short, giving you the basics, some activities to practise and references to other texts and websites.

Whole guides are written about essay writing, dissertations and fieldwork. Where you need more information, follow it up with the literature, online and at your university's skill centre.

Remember that there is a real difference between reading about a skill and applying the ideas in your degree. The **Try This** activities are designed to make the link between skills and their practical application, and to give you an opportunity to practise either mentally, or mentally and physically.

After the first two editions of this book there were requests for 2000-word examples of really badly written essays and hopeless PowerPoint presentations. There isn't room for either of these, but there are lots of changes in response to updating requests. There are more student and lecturers' feedback points included throughout.

I hope that you will enjoy some of the humour; this book is meant to be gneiss and fun with serious points. Geologists probably have the best jokes, but they are all pretty dire. The crosswords follow the quick crossword style familiar to readers of UK newspapers; the answers have vague GEES connections. The fun activities give your brain a chance to have a quick break, and keep you cheerful. Keep remembering to enjoy studying at university – it is exciting and fun, and a challenge.

KEEP SMILING!

ACKNOWLEDGEMENTS

She just doesn't get my geology jokes.

Many thanks to all the GEES undergraduates and post-graduates at the University of Leeds and University of Plymouth that survived the Study Skills module and contributed ideas and prompts for the book. Thanks are especially due to Mark Newcombe for graphics, Debbie Phillips, Martin Purvis, Andrea Jackson, Brian Chalkley, Sue Hawksworth, Sarah Underwood, David Bulmer, Sylvie Collins, Susie Stillwell, Michael Sanders, Yolande Knight, Jane Dalrymple, Sue Elm, Sarah Underwood, Glen Crust and all of the Geography, Earth and Environmental Science colleagues who have offered ideas and feedback on previous editions.

That's okay, igneous is bliss.

Thanks are due to Victor Gollancz Ltd for permission to quote extracts from *Interesting Times* (1994) and *Maskerade* (1995) by Terry Pratchett.

Study skills – why bother?

'Teach? No,' said Granny. 'Ain't got the patience for teaching.
But I might let you learn.'

(Pratchett, 1995, *Maskerade*)

This book discusses the skills that will help you to study effectively for degrees which involve Geography, Earth or Environmental Sciences (GEES). Some of the motivation for assembling it came from a student who said, 'The problem with first year was I didn't know what I didn't know, and even when I thought there was something I was supposed to know I didn't know what to do about it'. University can be confusing. It is not the same as high school. The biggest change is that you are expected to learn independently, rather than being taught. This book might help. It is deliberately 'hands-on', making lots of practical **Try This** suggestions. It aims to add to your self-confidence in your research and study abilities, and to save your time by acting as an ongoing resource. Rather than worrying about what will happen in a seminar, how to search online, or referencing in an essay, look it up and carry on. You are already skilled in thinking, listening, note-making, writing … BUT reviewing your approach and refining your skills should prove beneficial.

The language and tone of this book is deliberately light-hearted, with some games for light relief. There are some terrible jokes – keep smiling as you groan. Light relief is vital in study, but if you find deep thinking leads to deep kipping, have a coffee and solve a crossword clue – just remember to go back to thinking after your break!

Talk to friends and family who have been to university. Ask how they found it compared with school. Some answers are in Table 1.1 – what other insights do they add to this list?

This book aims to help you to notice how you learn and work, and to suggest ways of developing strategies that are effective for you. This should benefit you in the short term during your degree and in the longer term at work. Universities should be a good mix of fun, meeting lots of people and knowing more about the subjects you have chosen to study. What do your friends consider to be the benefits of their university experience? Their replies might include:

• Getting a graduate, better paid, more interesting job.

- Having the chance to do interesting things.
- Having a more interesting life.
- Getting a graduate, better paid, more interesting job.
- Having the chance to do interesting things.
- Having a more interesting life.

In my opinion the people who do well and get the most from university keep a good balance between study and fun. Most people want to perform at a high standard in all assessments to get a good degree classification, write assignments well because writing is a skill that employers want, be happy making presentations, be healthy and fit, and to develop good relationships with lots of people knowing that some will be lifelong friends.

An increasing number of students are doing GEES degrees without a previous high school background. Earth Science students usually studied sciences at school, but little geology. This is not a disadvantage: you are interested in the subject, all topics are fresh, and not confused by half-remembered notes (see Chapter 7, p.28) all the GEES degrees draw on many subjects for theory and insights, and some lecturers come from different disciplines. Your previous experience of history, statistics, economics, physics, sociology, mathematics, politics, chemistry and biology will all be useful at some stage. Mature students, with more experience of life, politics, business processes, social conditions and general knowledge, have an extensive skill base to build on.

University is not like High School	
Lectures have hundreds of people.	The classes are harder.
There are so many bright people.	You have to organize yourself all the time.
There were always so many other things to do, going out …	At the start you don't think it's going to be too difficult, but if you want to do well it is much harder than school.
Lots of people want you to do things with them.	Mum isn't there to tell you what to do.
Tutors expect you to talk about stuff … and think for yourself; it's not just about copying down the notes.	I really wasted my first year, it was fun but it made the second year much harder.

Table 1:1 University is not like high school

1.1 Independent learning – what does it mean?

University is different. The basic idea is that students are guided by academic staff to learn. Lecturers will help, but the responsibility for learning lies with each student. This is *independent* or *autonomous* learning. It allows people of different ages, backgrounds and interests to study together and graduate with degrees in the same subject, but each person will have explored his or her own unique combination of materials (Figure1:1).

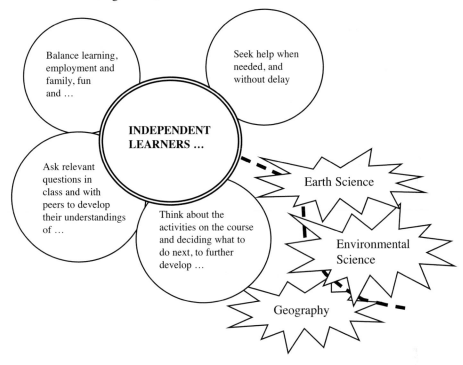

Figure 1:1 How independent are you?

Effectively university is about taking personal control of what you do and how you do it. There are modules, fieldwork, laboratories and time to explore many avenues. If, in the process, these equip you for later life, that is a bonus. GEES (geography, earth or environmental sciences) degrees have two elements:
- **The knowledge** – this will involve all the current theories, from cultural and medical geography, the racing speed of warm-based glaciers and fossil forests in Antarctica, to the consequences of deforestation, geopolitics, tectonics, sustainability and enterprise. The range and scope of GEES studies is planet-wide and deep.
- **The skills** – GEES graduates polish an excellent range of skills which have longer-term benefits in the workplace. Most GEES graduates will acquire through practical experience or osmosis the skills and attributes shown in Figure 1:2.

University is about learning how to learn. Most teaching involves active learning in environments where there is interaction, discussion and collaboration. Communication is crucial to renew and create ideas. Self-motivated learning is a vital skill

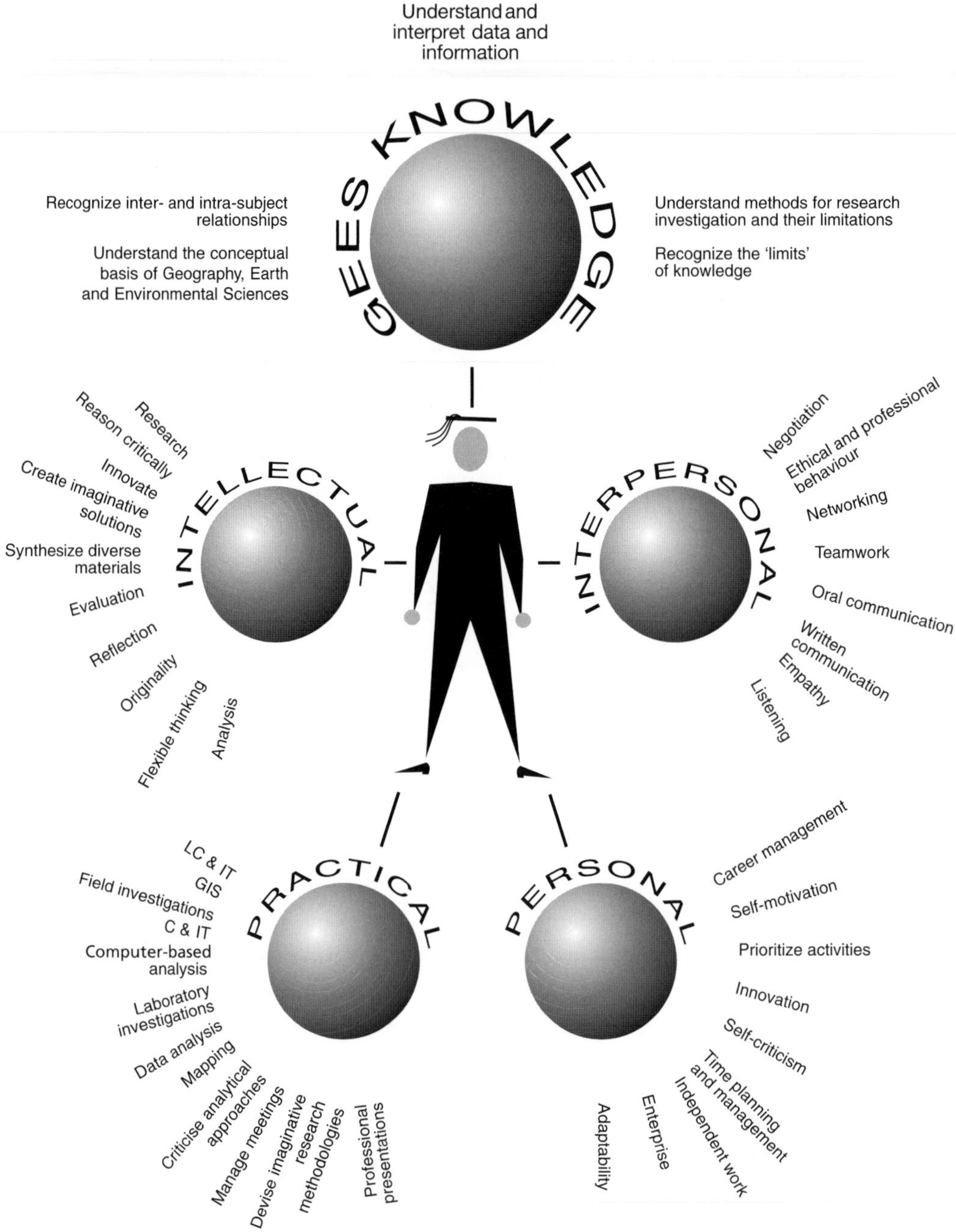

Understand and interpret data and information

GEES KNOWLEDGE

Recognize inter- and intra-subject relationships

Understand the conceptual basis of Geography, Earth and Environmental Sciences

Understand methods for research investigation and their limitations

Recognize the 'limits' of knowledge

INTELLECTUAL

Research
Reason critically
Innovate
Create imaginative solutions
Synthesize diverse materials
Evaluation
Reflection
Originality
Flexible thinking
Analysis

INTERPERSONAL

Negotiation
Ethical and professional behaviour
Networking
Teamwork
Oral communication
Written communication
Empathy
Listening

PRACTICAL

LC & IT
GIS
Field investigations
C & IT
Computer-based analysis
Laboratory investigations
Data analysis
Mapping
Criticise analytical approaches
Manage meetings
Devise imaginative research methodologies
Professional presentations

PERSONAL

Career management
Self-motivation
Prioritize activities
Innovation
Self-criticism
Time planning and management
Independent work
Enterprise
Adaptability

Figure 1:2 Skills and attributes of Geography, Earth and Environmental Sciences graduates

for life, enabling you to keep abreast of developments and initiatives. Graduates from the 1960s, before computers, have had to get to grips with technology while at work; what will happen in the next 25 years? Employment is unpredictable. Job markets and business requirements change rapidly. Employers need individuals who are thoughtful and flexible about their careers. An effective graduate is someone who sees their career as a process of work and learning, mixing them to extend their skills and experience. This is the essence of lifelong learning, and university is part of it.

To add value to your degree, it helps to think about what you do every day at university (see Chapter 2), to give you skills that have market value. Employers claim to be happy with the academic skills students acquire, such as researching, collating and synthesizing new material, but they also want graduates with skills like listening, negotiating and presenting. Any strengthening of your skills and experience of skill-based activities should add to your self-confidence and improve your performance as a student and as a potential employee. Expect to be involved in exercises and activities that include:

- Research
- Practical work
- Fieldwork
- Role-plays
- Negotiation
- Interviews
- Self-guided study
- Learning contracts
- Decision-making
- Personal development planning
- Action planning
- Simulations
- Case studies
- Work placements
- Self-evaluation and review
- Group work
- Time management
- Presentations.

Your degree will give you the opportunity to experience the latest e-action, wikis, blogs, podcasts, vodcasts, electronic books and journals, video-conferencing, e-mail, spreadsheets, e-learning, and Virtual Learning Environments (VLEs). The technology may seem daunting but it is fun too. (And if some five-year-old proto-anorak wearer can manage, so can you!)

University teachers know that effective graduates are comfortable and confident with knowledge from their degrees (petrology, economics, air pollution) and the ways in which they acquire and share that knowledge (group discussions, seminars, presentations, posters, reports). Good teachers encourage you to build on your experience through class activities which involve working with others as much as on your own, because the reality of work and research is that it's a team game. The skills that you will develop during your degree include:

✓ communication skills, both written and oral, and the ability to listen to others;
✓ interpersonal or social skills: the capacity to establish good, professional working relationships with clients and colleagues;

✓ organizational skills: planning ahead, meeting deadlines, managing yourself and co-ordinating others;

✓ problem analysis and solution: the ability to identify key issues, reconcile conflicts, devise workable solutions, be clear and logical in thinking, prioritize and work under pressure;

✓ intellect: judged by how effectively you translate your ideas into action;

✓ leadership: many graduates eventually reach senior positions, managing and leading people;

✓ teamwork: working effectively in formal and informal teams;

✓ adaptability: being able to initiate and respond to changing circumstances, and to continue to develop one's knowledge, interests and attitudes to adapt to changing demands;

✓ technical capability: the capacity to acquire appropriate technical skills, including scheduling, information technology (IT), statistics, computing, data analysis, and to update these as appropriate;

✓ achievement: the ability to set and achieve goals for oneself and for others, to keep an organization developing.

OK, this is a long list. Lecturers focus on different things with their classes. Well-planned degrees have a good mix of all of these activities across the whole of the programme. By the time you graduate you should feel confident in listing all these skills and more on your curriculum vitae (CV), and be able to explain where in your degree these abilities were practised (Table 1:1).

Graduate skills	In GEES degrees
Learning how to learn	All modules. Taking personal responsibility for your learning as an individual and in group research for fieldwork, laboratory work, projects and dissertations.
Communication	All modules. Presentations in seminars, tutorials, workshops, debates, practicals and all written assignments.
Information Technology	Most modules will involve online research activities, data and word-processing, graphs, statistical analysis, GIS and programming.
Numeracy	All modules involving statistics, data handling, analysis of data in practicals, fieldwork, projects and dissertations.

Table 1:2 Where to find skills in GEES degrees

1.2 What to expect, and spot the skills!

Geography, Earth and Environmental Sciences degrees are traditionally divided into three years called either Years 1, 2 and 3, or Levels 1, 2 and 3. An additional year may be intercalated for an industrial placement or a year abroad. A year is typically divided into 10 or 12 teaching modules or units for 360 credits, addressing a range of topics. GEES degrees are described as progressive, which means that the standards and difficulty increase each year. Expect modules at Levels 2 and 3 to build on your Learning at Levels 1 and 2. This section outlines the main university activities and some of the skills they enhance.

Lectures

Believing any of the following statements will seriously damage your learning from lectures:

- In good lectures, the lecturer speaks, the audience takes very rapid notes and silence reigns.
- The success of a lecture is all down to the lecturer.
- A great lecturer speaks slowly so students can take beautifully written, verbatim notes.
- Everything you need to know to get a first class degree will be mentioned in a lecture.
- Lectures are attended by students who work alone.

Lectures are the traditional teaching method, usually about 50 minutes long, with one lecturer and loads of students. Lectures involving 300+ students can seem impersonal and asking questions is difficult. A 10-credit module will probably have 20 hours of lectures but this is just part of the 120 hours you are expected to spend studying the topic. The lecturer is aiming to introduce the topic in a way that encourages you to rush to read in much more depth and really understand the topics that interest you. Good wheezes to manage lectures include:

☺ Download and read the PowerPoints for the lecture from the VLE. In many universities they are available up to a week ahead.

☺ Get to the Lecture Theatre early and find a seat where you can see and hear.

☺ Have a supply of paper, pens and pencils ready.

☺ Get your brain in gear by thinking, 'I know I will enjoy this lecture, it will be good, I really want to know about ...'; 'Last week s/he discussed ..., now I want to find out about ...'

☺ Before the lecture, read the notes from the last session, and maybe some library material too. Skim the PowerPoints. Even 5–10 minutes will get the old brain in gear. If you know roughly what is being covered, you remember more. Making notes on PowerPoints saves you writing down the main points – you can add other details.

☺ Look at handouts carefully. Many lecturers give summary sheets with lecture outlines, main points, diagrams and reading. Use these to plan reading, revision and preparation for the next session.

☺ Think critically about the material presented. What did you not understand? What do you need to read to get more case examples? What is interesting? Why is this topic important for geography, earth or environmental science?

☺ Revise and summarize notes soon after a lecture. It helps revision later on and you can see where there are gaps in your understanding, which helps you to decide what to read.

☺ Talk about the topic with friends, study buddies and others.

Skills acquired during lectures include understanding GEES issues, recognizing research frontiers and subject limitations.

Tutorials: What are they? What do you do?

Usually tutorials are a 50-minute discussion meeting with an academic or post-graduate chairperson and 4–8 students. The style varies between departments, but there is normally a topic involving preparation. You might prepare a short talk, an essay, an outline essay, material for a debate, review a paper, produce a short computer program, and share your information with the group. The aim is to discuss and evaluate issues in a group that is small enough for everyone to take part. Other jolly tutorial activities include brainstorming examination answers, working through maths or statistics problems, comparing note styles, creating a poster, planning a research strategy, discussing the practicalities of a fieldwork proposal, evaluating dissertation possibilities, and the list goes on …

The tutor's role is NOT to talk all the time, NOT to teach and NOT to dominate the discussion. A good tutor will set the topic and style for the session well in advance, so everyone knows what they are doing. S/he will let a discussion flow, watch the time, make sure everyone gets a fair share of the conversation, assist when the group is stuck, and sum up if there is no summarizer to do so as part of the assignment. A good tutor will comment on your activities, but tutorials are YOUR time.

> I don't really like tutorials because they make you think. You cannot just sit there. I do know that listening and talking about stuff makes me think, and that gives me ideas.

Some tutors will ask you to run a couple of tutorials in their absence, and to report a summary of the outcomes. This is not because tutors are lazy, but because generating independence is an important part of university training. Student-led and student-managed tutorials demonstrate skills in the management of group and personal work. When a tutor is ill, working unsupervised uses the normal tutorial time effectively. A tutor may assign, or ask for, volunteer chairpersons, timekeepers and reporters to manage and document discussions.

Your role is to arrive at tutorials fully prepared to discuss the topic NO MATTER

HOW UNINTERESTED YOU ARE. Use tutorials to develop listening and discussion skills, become familiar with talking around GEES issues and build up experience of arguing about ideas.

➡ **TOP TIPS**

➜ Taking time to prepare for tutorials will stop (or reduce) nerves, and you will learn more by understanding a little about the topic in advance.

➜ Reviewing notes and reading related material will increase your confidence in discussions.

➜ Asking questions is a good way of saying something without having to know the answer.

➜ Prepare a couple of questions or points in advance and use them early on, get involved.

➜ Taking notes in tutorials is vital. Other people's views, especially when different from your own, broaden your ideas about a topic, but they are impossible to recall later unless noted at the time. Tutorial notes make good revision material.

Tutorial skills include communication, presentation, critical reasoning, analysis, synthesis, networking and negotiation.

Seminars

Seminars are a slightly more formal version of a tutorial, with 8–25 people. One or more people make a short presentation, leaving ample time for group discussion. Seminars are a great opportunity to brainstorm, to note the ideas and attitudes of colleagues, to spot extra examples and approaches. Take notes.

Even if you are not a main speaker, you need to prepare in advance. In the week you speak, you will be enormously grateful to everyone who contributes to the discussion. To benefit from this kind of co-operation, you need to prepare and contribute in the weeks when you are not the main presenter.

ONLY TO BE READ BY THE NERVOUS. (Thank you.) If you are worried, nervous or terrified, then volunteer to do an early seminar. It gets it out of the way before someone else does something brilliant (well, moderately reasonable) and upsets you! Acquiring and strengthening skills builds your confidence so seminars appear less of a nightmare. By the third week you will know people and be less worried.

Seminar skills include discussion, listening, analysis, teamwork, giving a professional performance and networking to make more connections than Facebook.

Workshops or large group tutorials

Workshops (large group tutorials, support or revision classes) are sessions with 12–35 students that support lecture and practical modules. They have very varied formats. There will usually be preparation work and a group activity. Tutors act as facilitators, not as teachers. Expect a tutor to break large groups into subgroups for brainstorming and discussions. These sessions present a great opportunity to widen your circle of friends and find colleagues with similar and/or diverse views, *and* develop discussion, argument and listening skills.

Workshop skills are the same as for tutorials and seminars, with wider networking and listening opportunities.

Computer and laboratory practicals

Practical classes and fieldwork are the 'hands-on' skills element of all GEES degrees. Many departments assign practical class time, when tutor support is available, BUT completing exercises and developing your proficiency in IT, computing and laboratory skills will take additional time. Check the opening hours of computer laboratories on campus.

In laboratory practicals, always take note of safety advice, wear lab coats and safety glasses as advised, and please don't mix acids without supervision. Most laboratory staff are trained in first aid, but would rather not have to practise on you.

Assessment

Assessment comes like Christmas presents, regularly and in all sorts of shapes. Assessments should be regarded as helpful, because they develop your understanding. There are two forms:

✓ Within module assessment of progress, where the marks do not count (formative), and usually involves lots of feedback.

✓ Assessments where the marks do count (summative). Feedback styles vary. The results eventually appear on your degree result notification for the edification of your first employer who wants written confirmation of your university prowess.

> *Isn't the skills part of it just the stuff we had to do at school again?*
>
> No, some of the topics will the same (presentations etc.), but at uni it is focused on behaving like a professional, developing a calm, polished and confident approach in all situations …

There is a slight tendency for the average student to pay less attention to formative, within module assessments, where the marks do not count. Staff design formative tests because they know 99 per cent (± 1 per cent) of students need an opportunity to 'relax and have a go', to understand procedures and what is expected, because the marks do not matter.

Assessments are very varied: examinations, essays, oral presentations, seminars, posters, discussion contributions, debates, reports, reviews of books and papers, project designs, critical learning log, fieldwork notebooks, laboratory competence, computer-based practicals, multiple choice tests ... It is a matter of time management to organize your life and them (Chapter 4). You should know in advance exactly how each module is assessed and what each element is worth. Many modules have mixed assessments, so those who do very well in examinations or essay writing are not consistently advantaged. Many departments have assessment and feedback criteria. Get hold of your own departmental versions or see examples for essays (Figure 17:2), oral presentations (Figure 13:3), practical reports (Figure 18:2), dissertations (Figure 21:2) and poster presentations (Figure 26:3). If you cover all the criteria then the marks come rolling in.

> Lecturers use peer assessment to help students develop their critical assessment skills and become more comfortable with giving and receiving feedback.

Amongst the many skills enhanced by assessments are thinking, synthesis, evaluation, originality and communication.

Feedback

Students receive, and give, feedback all the time; from tutors, lecturers, study buddies, friends, and everyone else who has their interests at heart. Lecturers provide feedback because they want all their students to learn effectively, whereas students tend to want feedback that explains why they achieved a particular mark. The trick is to understand that lecturers use feedback as a form of coaching. They use many different styles and approaches to encourage their classes.

Use **Try This 1.1** to make notes about how you respond to feedback positively (*great idea, doing it now*), and negatively (*don't you tell me what to do*) and how you can make the most of the advice offered.

Advice, comments, thoughts, information, opinions, suggestions, guidance, guidelines, instructions, recommendations, views, perspectives ... these are ALL forms of feedback. Remember people aim to help. Feedback is designed to enhance your skills and performance. Comments are made about your work, not about you as a person.

Where classes are very large feedback may happen some weeks after you finished an assignment. It is too easy to glance at the mark and ignore your tutor's detailed comments. Work out a strategy to make sure you benefit from this feedback to improve your next assignment.

⚙ **TRY THIS** 1.1 – Handling feedback

Feedback arrives in many ways. How will you respond to each style to make the most of the advice? Make some notes.

Verbal feedback from a lecturer in a large class/lecture.	Comments and ideas from a study buddy.	Text message from a tutor.
Online discussion using Skype, tutor to student, and students in peer groups.	Written feedback on a report or essay.	Verbal feedback podcast sent by mobile phone and VLE.
Conversation with your tutor in the corridor, refectory or gym.	Videocast from the marker of an exam, arrives three weeks later.	Summary written feedback on a class exercise online in the VLE/handed out in class.
Advice in module handbooks. This may include feedback on student's work in previous years.	Peer tutors (students mentoring students).	Information from students in the year above you. (It's all feedback.)

Handling feedback so that you learn from it is a major university skill that has real workplace value. What makes you act on feedback? What motivates you to act?

Every idea you have about what you do is a form of feedback to yourself.

VLEs

Most universities have a Virtual Learning Environment (VLE), although it may be called something else such as BlackBoard or Moodle. The VLE gives you access to module and time table information, lecture notes, PowerPoints, the library, assessment information and your exam results. It's where module leaders organize online discussion groups, put answers to questions students ask, and … whatever it is called, you will save oodles of time by doing the online tutorial and knowing your way around the system quickly. Being familiar with how your VLE works will make your life much easier.

Non-academic learning

Do not underestimate what you know! In your years at university you acquire loads of personal skills, like negotiating with landlords, debt crisis management, charming bank staff, juggling time to keep a term-time job and delivering essays to deadline, being flexible over who does the washing up, and handling flatmates and tutors.

You get feedback from everyone. People talk to you about stuff, tutors, mates, the people in hall. Feedback from peers and supervisors improves what you are doing. I had to work out how to use it.

1.3 The research process

All students are 'reading' for a degree, finding out for themselves about research processes. Taught activities involve 20–50 per cent of the timetable, leaving 50–80 per cent for your own research into sub-topics through activities that reinforce your understanding. The scope of GEES topics is vast, certainly global and at times interplanetary. Covering all aspects is impossible. Your challenge is to develop your interests and knowledge through researching different elements of the topics involved in your GEES degree, within the constraints of time, facilities and energy. Consequently, researching in a group (see Chapter 15) can be seriously beneficial.

The issues addressed in the GEES degrees are not simple. Lecturers will indicate what is already well understood and discuss areas of the subject where we are less sure about what happens, pointing out where knowledge is missing, provisional, uncertain and worthy of further investigation. Most issues are interlocking and multidimensional. By the time you graduate you should be able to take complex, unclear and, at times, contradictory information from a wide range of sources and synthesize it to make sense of the picture at a range of scales. For a number of topics which you have studied in depth you should have an enhanced ability to recognize both the boundaries of knowledge, what is known and what is not known, and what you as an individual know and do not know. Recognizing the boundaries of one's own expertise is a relevant life skill. Someone who does not understand the implications of their actions in changing procedures, for example, is a potential danger to themselves and the wider community.

University learning is not about recalling a full set of lecture notes. It is about understanding issues and being able to relate and apply them in different contexts.

1.4 How to use this book

No single idea is going to make a magic difference to your learning, but taking time to think about the way you approach tasks like reading and thinking, listening and writing, researching and presenting should help your efficiency rate. Studying is a personal activity. There are no 'right' ways, but there are tips, techniques, short cuts and long cuts.

Attempting to read this book in one go will not help. Look through the chapter headings and index. Read a couple of things that interest you now. When you are worried or stuck, then, hopefully, there is a useful section. Use this book it as a guide throughout your degree. Some parts are relevant for level 1, others, like the dissertation advice (Chapter 21), will matter more in the last year. When you have an essay to write look at that chapter. No one expects you to know the whole of the

Greek alphabet or all the Latin names for plants, but if someone mentions it, there is a bit of this book (28.6 and 28.7, p.277-9) that will help.

> Take feedback seriously, think carefully. If you disagree with some of it, it may be best to discuss why with your tutor. When advice is balanced and constructive it is probably useful. Discuss feedback with your friends and tutor.

There are lots of **Try This** activities to encourage you to get involved and build on your existing skills. Adapt the activities to your needs. Some statements in the book are deliberately controversial, designed to encourage thinking. Most of the figures and examples are deliberately 'less than perfect'. Consider how they can be improved; it's called active criticism. Universities have IT and sports facilities, getting more skilled means using them and 'working out'. Your first study year is a good time to practise and enhance learning skills as you adjust to your new life, but it is important to keep practising and reflecting throughout your degree. Experience is built by doing, not by watching.

1.5 References

There are many generic skills texts – check out the library.

Gregory KJ, Simmons IG, Brazel AJ, Day JW, Keller EA, Sylvester AG and Yanez-Aranciba A 2009 *Environmental Sciences: A student's companion*, Sage, London

Rogers A and Viles H (Eds.) 2003 *The Student's Companion to Geography*, 2nd Edn., Blackwell, Oxford

Geograms 1

Reorganize the letters to find a GEES-related term. Answers on p 295.

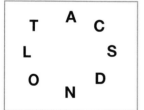

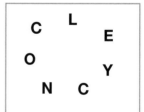

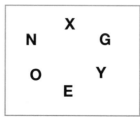

2

Reflection skills, reviewing and evaluating, adding value to your degree

Experience is something you don't get until just after you need it.

Degrees involve personal decisions about what to read, research, ignore, practise, panic over ... You may do exactly the same modules as 100 other students, but you will learn different things. Reviewing what you are doing reduces worries and provides some rational options when there are choices. Taking control and responsibility for your work and acting on feedback can seem scary. The lack of guidance about academic work is one reason why many new university students feel disoriented, chucked in at the deep end without a lifeguard in sight. Actively managing yourself really helps to cope with submission deadlines.

All UK universities have processes to help students to plan and reflect on their progress during their degree. There is a structured format for planning and a record-keeping system. Records may be kept in a paper-based portfolio or log book, but are more likely to be held in an e-portfolio operating through the university's VLE (Virtual Learning Environment) or a website. The systems have various names usually involving a combination of the following words: personal development, planner, portfolio, e-portfolio, progress file, learning-log and reflection. PDP (personal development planning) is a regular acronym. The advantage of e-portfolio PDP systems is that you can easily import pictures and video from other sources such as Flickr or YouTube, and examples of work from word-processing and database files. During group work you can share feedback and thoughts more easily than in paper-based systems. Your programme will introduce you to details of your own university's system, but if you are reading this before university and want to know more, look at the online PebblePad (2011) overview slides.

This chapter briefly discusses why evaluation and reflection are useful skills, and how reflection techniques help decision-making in your studies. See the tutorials and support materials for your own university's PDP system. There are a variety of **Try This** activities because everyone has their own needs and priorities – different activities are relevant at different times of the year and in different years of your degree. Most activities benefit from a 'mental and physical' (pen in hand, fingers on the keyboard) approach. The knack is to develop reflective thinking so that it becomes automatic; reviewing becomes a process done while warming up at kick boxing or cleaning out the hamster cage.

The skill benefits from this chapter are thinking, evaluation, reviewing and reflection. Reflection skills are not learned easily or acquired overnight. They develop with practice; the result of thinking actively about experiences and placing them in a personal context. This is an iterative process.

2.1 Why reflect?

Sometimes you have to step back to leap forward.

Adding value to your degree

Some students find it hard to see how parts of their degree course interconnect.

Taking a little time to think about interconnections between modules, 'why economic costing and pricing principles are important for cultural geographers or flood forecasters', or 'where statistical tests can be used in your dissertation', can give your modules more cohesion and be motivating.

For most of my degree we did lots of different modules and they were interesting in a way, but I couldn't see why we were doing economic geography and then statistics and the resources bits and so on. I couldn't really see the point of working at the bits. It was about halfway through my final year when all the bits began to make sense and slot together. And then it got to be really interesting and I could see why I should be doing more reading and I did quite well in the last semester.

Increasing employability

Many employers look for enthusiastic graduates with skills of articulation and reflection, those who can explain, with examples, and evaluate their experience and qualities. Recruiters want to identify people with the awareness and self-motivation to be proactive about their learning. The ability to teach oneself, to be aware of the need to update one's personal and professional expertise, and to retrain, is vital for effective company or organizational performance and competition.

Most large organizations have some form of professional PDP process as part of their staff development and appraisal. Keeping accurate records is critical for professional standards in a range of jobs. Earth Scientists, technologists and water engineers will have professional body e-portfolios to complete throughout their careers. Where possible, university portfolios mirror your professional body portfolio (e.g. Geological Society 2011, CIWEM 2011).

> *The lectures were OK but the pracs seemed really tedious. There was loads to do ... When we got to the project work, the lecturer assumed we could do the analysis because we had had the pracs. It was obviously really useful then. ... Lecturers tell you stuff will help with dissertations and projects but you don't really believe them until you have to use the stuff later.*

Your academic background may be of little interest to an employer. Whether you are expert in modelling the spread of disease, have researched Mongolian housing patterns or abseiled down a glacier is not important. What is relevant is that faced with the task of researching the market for a new type of chocolate, you can apply the associated skills and experience gained through researching the nineteenth-century development of the Co-operative Movement in Rochdale, or new waste control management processes (thinking, reading, researching, presentation, making connections) to designing and marketing cocoa products. It is your ability to apply the skills acquired through school and university in a workplace role that employers value.

> *Filling in the portfolio just seemed a crazy waste of time at Uni, but you just do it every day at work and it's so normal. It's how you keep track of everything you do.*

Remember that an employer is looking for a mix of skills, evidence of your intellectual, operational or practical and interpersonal skills (look back at Figure 1:2). Your GEES intellectual skills are demonstrated by your degree certificate. Keeping a record of your thoughts in your e-portfolio or on forms like those in the **Try This** activities here, or in a diary or journal-style log, pays off when filling in application forms. They will remind you of what you did and of the skills involved.

If you plan to work in very large companies where there are many training courses, your lifelong learning will be enhanced by company training. If you want to establish your own business or join small and medium-sized enterprises (SMEs), with small numbers of employees and budgets, then your university-acquired reflection skills will be directly beneficial.

2.2 Getting started

Some businesses require new staff to keep a daily log in their early years of employment. It encourages people to assess the relative importance of tasks and to be efficient managers of their time. It is a reflective exercise in which, at 4.50 p.m. each day, you complete a statement like:

I have contributed to the organization's success/profits today by ....

....

I was fully skilled to do ....

I was less capable at ....

Other comments ....

At the end of the week or month, these statements are used to prioritize business planning and one's Continuing Professional Development (CPD). It is an activity that most new employees hate. However, most admit later, that it taught them an enormous amount about their time and personal management style, and wished they had started sooner. In time, this type of structured self-reflection becomes automatic, individuals continually evaluate their personal performance and respond accordingly. The GEES student equivalent is:

I contributed to my degree today by ....

I could have been more efficient at doing this if ....

Tomorrow I am going to ....

There are a number of **Try This** activities in this chapter, each suited to different stages on a degree course. These can be used to build up a learning log, recording your university experience. A learning log can be just a diary, somewhere to note activities and skills; a reflective log asks for a more detailed, reasoned response.

My first year tutor got us to fill in the forms, and you do get better by doing the practicals and some exercises. My second year tutor made everyone pick a skill to focus on each term. Having decided to get better at chairing, I volunteered to chair each of our group work activities and I wouldn't have done it if he hadn't made me pick one thing. It felt crazy weird because I don't normally volunteer but it was good, and I liked it after a bit.

2.3 Reflecting on your degree skills

Try This 2.1 asks you to articulate your feelings about your current personal approach to learning and your degree course. If you evaluate your response and act on it, you are taking charge.

> *It does make you more organized and independent.*

⚙ TRY THIS 2.1 – Initial reflections

Here are some example reflections. What are your thoughts? Make some notes as you think.

Skill	Reflections
Speaking in tutorials and seminars.	*I know I don't say enough in tutorial. I know what I want to say but it all seems so obvious I feel silly, so I guess I need to get stuck in early.* *PLAN: to answer first, learning log can act just as a diary, somewhere to note activities.*
Knowing when to stop reading.	*I read up to the last minute, and then rush the writing.* *PLAN: to put reading deadlines in my timetable.*
Including relevant information in essays.	*I try to include everything in an essay to show I have done some reading.* *PLAN: be more selective, somehow, next time.*
Organising ideas coherently.	*I know I can do this if I plan an essay properly, getting paragraph ideas in order.* *PLAN: just do it next time.*
Being more open to new ideas.	*PLAN: buy a newspaper. I could take notes in tutorial. More reading might be okay!*
Making time to sit down and think about different ideas.	*This seems really odd, because you sort of do thinking all the time. It's not really a cool activity. Could try when no one knows – in bed maybe?*

What are your thoughts on:
Making notes
Listening carefully in discussion and responding
Delivering essays on time
Using diagrams to illustrate essay ideas
Drawing the thread of an argument together to developing a logical conclusion
Negotiating
Putting ideas into my own words
Disagreeing in the discussion without causing upset/being upset
Reading more widely?

Try This 2.1 is a self-assessment exercise that you might want to repeat after a term or semester. It is useful to remember that when people self-assess a skill before and after an activity, the assessment at the end is frequently lower than that at the start.

Although the skill has been used and improved during the activity, by the end, it is possible to see how further practice and experience will lead to a higher skill or competence level. Now look at **Try This 2.2**. In thinking about your strengths and weaknesses, talk to family and friends, ask what skills you have and what you do well.

TRY THIS 2.2 – Being active about skill development

Having completed **Try This 2.1** go back to the list and highlight three skills you would like to act on in the next three months. Make some notes about what to do about these three issues, add some deadline dates. Like New Year resolutions, this activity needs revisiting.

How do you go about making decisions? (See **Try This 2.3**.) Which of the following characterizes your approach? You will probably ✓ a number depending on the circumstances, but overall are you in control of you? Are you making your own decisions? What action will you take?

TRY THIS 2.3 – Who controls your life?

Consider the decisions you have made in the last week and last five years. Who really made the decisions? Which of these processes do you tend to adopt? Are you happy with your approach?

Hopeful	Choosing the option that should bring a happy result.
Go for it	Get straight on with the first option, without considering other paths.
Tomorrow will do	Leaving decisions until well past critical times, putting things off.
Alternatives are overwhelming	Researching thoroughly, getting so much information that you cannot decide on priorities. (Cannot see the wood for the trees.)
Following the crowd	Letting the group or another individual do the thinking and deciding for you.
Fatalistic	Letting life happen, not being prepared for potential eventualities.
Missing out/Avoidance	Taking yourself elsewhere so that your imagined 'worst case scenario' cannot happen.
Risk-averse	Taking the safe, best-odds opportunities.
Sorted	Having a systematic, logical route to decide what to do.
Psychic	'The aura is good', 'This feels okay' style of decision making.

TOP TIP

→ Reflection is better and reinforced when you write down your thoughts or speak them aloud.

2.4 Within module reflection

If reviewing the day or term is too much of a drag, think about your modules. On field class, as part of laboratory or dissertation work, PDPs act as both diary and reflective statement. You can use the same process to help you think about other modules. If you find lecturers use the PDP process in first year modules, but not in second- or third-year modules, it's probably because they are familiarising you with the processes in first year so that you will use them automatically later in your degree.

> Students summed up a PDP workshop session as: *Inspirational; Fun; Useful; Highlights importance of teamwork; Difficult; Confusing; Captures your imagination; Makes you think on your feet; Disorganized; I don't have to be in a mess; Made me think about something I'd never considered before; Can I wait until I get to work?*

Students

Variations on your responses to **Try This 2.4** will answer questions asked in interviews. Thinking about how you work in advance, and getting used to talking and writing about it, gives you more chance to enthuse and be positive. You may not have had much experience of some skills, but any experience is better than none.

Sometimes one recognizes that a particular lecture, laboratory class or day has passed without being of any real benefit to one's degree! Try to identify why. **Try This 2.5** and **Try This 2.6** contain some ideas.

TRY THIS 2.4 – Using PDP reflection in modules

Read the module handbook and learning outcomes. List the skills you expect to use and develop.

Pick one or two you want to focus on *actively* developing. (Doing the module should develop all the skills, this is about choosing particular areas to focus on.) Create a brief action plan to achieve each goal. Get started. Make some notes about how you are developing every two weeks.

Draft some answers as mini case examples that you could use at an interview to support your statements. See p 295 for some students' thoughts.
- What I want to get out of attending this module is ...
- This module/session helped me to develop a clearer idea of my strengths and weaknesses, for example ...
- I have discovered the following about myself with respect to decision-making ...
- The skills I used well were ... Other members of the group showed me new ways to ...
- The preparation for my (the group) presentation was ... My points to improve next time are ...
- Our group could have done better if ...
- I (our group) made decisions by ... My role was ...
- I have learned ... about interview technique/asking questions/planning laboratory work/ investigating in the field.
- I most enjoyed ... about the exercise/session/module/degree course.
- I least enjoyed ... about the exercise/session/module/degree course. Next time I plan to ...
- The biggest challenge to me in this exercise/session/module/degree was ...

Start by using **Try This 2.4** for one module, it will take a few minutes. Pick a module you're doing now and take ten minutes. Remember the module handbook is on the VLE/on line/ ... in your files.

TRY THIS 2.5 – Reflecting on a day

Brainstorm a list of things that happened (5 minutes maximum, just a back-of-an-envelope list) e.g.:
Went shopping.
Had hair cut.
Went to Dr Impossible's lecture.
Talked to Andy all afternoon.

Then brainstorm a list of things that made the class/day unsatisfactory, e.g.:
Andy talked to me for hours.
Bus was late.
I didn't understand what Dr Impossible was going on about.
Printer queues were hours long.

Leave the two lists on one side for a couple of hours. Then grab a cup of coffee, a pen, reread your lists and make a note about where you might have saved time, or done something differently. Consider what might make life more satisfactory if these situations happen again:
Natter to Andy for an hour MAXIMUM! over coffee and leave.
Take a book on the bus.
Have a look at Dr Impossible's last three lectures. If it still doesn't make sense I will ask my tutor or Dr Impossible.
Need to take something to read, or do online searches, while print-outs are chugging through.

TRY THIS 2.6 – Reflecting on an activity

What have I learned from doing this essay/presentation/report/mapping project ...?

What worked least well? Why did it not go so well? What could I do next time to make this easier?

How have I developed my ideas in this essay/project/...?

Was it worth the effort that I made? Did I have a good balance of reading/thinking/writing/revising/finalizing?

How has the feedback I got helped?

How does this piece of work help me in preparing for future assignments and exams?

What have I learned from doing this essay/presentation/report/mapping project ...?

What worked least well? Why did it not go so well? What could I do next time to make this easier?

How have I developed my ideas in this essay/project/...?

Was it worth the effort that I made? Did I have a good balance of reading/thinking/writing/revising/finalizing?

How has the feedback I got helped?

How does this piece of work help me in preparing for future assignments and exams?

2.5 What to do first?

There are many competing demands on your time, and it is not always obvious whether the next research activity involves finishing a practical report, browsing Google Scholar or the library shelf for next week's essay, or reading another paper. Reflect on who or what takes most of your time. Some tasks do take longer than others, but the proportions should be roughly right. Questions which encourage prioritizing tasks include:

- Why am I doing this now? Is it an urgent task?
- Is the time allocated to a task matched by the reward? For example, it is worth considering whether a module essay worth 50 per cent deserves five times the time devoted to a GIS practical worth 10 per cent?
- When and where do I work best? Am I taking advantage of times when my brain is in gear?
- How long have I spent on this web search/seminar preparation/mapping practical/Africa essay? Were these times in proportion? Which elements deserve more time?
- Am I being interrupted when I am working? If I worked somewhere else would that help?
- Who causes me to take time out? Are there ways of limiting this by say an hour a week?

> Don't take negative feedback personally. People are commenting on your work, not on you. They are taking time to help you to improve.

2.6 Start on your CV now

Use reflective material to amplify your CV. The thought of leaving university, applying for jobs and starting a career is probably as far from your mind as the state of the Bolivian economy. Nevertheless, for those desperate for money and applying for summer jobs, a focused CV can significantly increase the chance of selection for that highly paid shelf-stacker or burger-bar job. Reflecting on your skills at an early stage may highlight the absence of a particular 'skill'. There is time to get involved in something that will demonstrate you possess that skill, before the end of your degree. Have a go at **Try This 2.7** and check out Chapter 27.

> When we first talked about it our PDPs, the group agreed you just made it up and that is kind of what happens at school and here. You do it just for the teacher. Talking to Elle, you could see her line manager is the person she works with all day and talks to every day, so making it up isn't an option. And she was really positive about it helping her with doing a better job.

If you have forgotten what skills your modules involved, look back at the course outline. It is likely to include a statement like: 'On completion of the module students will have ...' Use these statements to amplify your CV and jog your memory.

> The session reminded me to start thinking about getting some work experience this summer, that I need to do something now.

Try This 2.7 does not include skills acquired through leisure pursuits or work experience. Compile a second list from those experiences. Driving, shorthand, stocktaking, flying, language skills, writing for a newspaper or magazine, treasurer, secretary and chair of societies all involve skills such as time management, negotiation, listening, writing reports and many more. Work experience does not have to be paid work; voluntary activities can give you valuable experience that pays dividends on a CV.

TRY THIS **2.7 – Skills from MY geography/environmental science/earth science degree**

Expand and tailor this list for your degree, from your university. Be explicit in articulating the skills and the evidence. Update it each semester. There are a few starter suggestions in the second column. **Skills acquired from MY**......................................**degree include:**

Numeracy	*Statistics modules in Years 1 and 2. Calculations for science laboratory experiments. I completed a financial balance sheet for a set of laboratory experiments and for my dissertation.*
Able to meet deadlines – essays, reports, practical write-ups, etc.	*All essays completed in time. Organized a group project and planned the mini deadlines that kept us on track.*
Organizational skills	*Final year dissertation, organized personal fieldwork in nature reserve, this required co-ordination with landowners, wardens and with the laboratory staff for analytical facilities.*

What would you say about your:

Teamwork skills
Workshop group work skills
Ability to put ideas across
Ability to work individually
Time management skills
Ability to prioritize tasks

Problem-solving experiences
Self-motivation
IT skills
Computing skills
Communication and presentation skills?

2.7 References and resources

Bournemouth University 2011 Welcome to Bournemouth University PDP, http://pdp.bournemouth.ac.uk/   Accessed 15 February 2011

CIWEM 2011 Continuing Professional Development, http://www.ciwem.org/membership/cpd.aspx   Accessed 15 February 2011

Geological Society 2011 Continuing Professional Development (CPD) & Training http://www.geolsoc.org.uk/gsl/op/www.geolsoc.or%3C/education/cpd   Accessed 15 February 2011

Imperial College 2011 Imperial College Employability, Careers, Imperial College London, http://www.imperial.ac.uk/ice/   Accessed 15 February 2011

Karimjee R 2011 An introduction to PDP, http://www.city.ac.uk/ldc/resources/Personal%20Development%20Planning/An%20introduction%20to%20PDP.html (podcast and information)   Accessed 15 February 2011

PebblePad 2011 Welcome to PebblePad, http://www.pebblepad.co.uk/   Accessed 15 February 2011

University of Cambridge 2010 What is Personal Development Planning (PDP)? Cambridge University Skills Portal, http://www.skills.cam.ac.uk/undergrads/pdp/planning/recording.html   Accessed 15 February 2011

Keywords for researching reflection: PDP, personal development planning, reflection, career, lifelong learning, graduate skills, volunteering, internships, career development.

Quick crossword 1

Answers on p 296.

Across
1 Smart conifer (6)
5 Period (4)
8 Young sheep (4)
9 Mine waste (8)
10 Dancer and Prancer (8)
11 Bog fuel (4)
12 Ouse and Trent estuary (6)
14 Running water (6)
16 Cast an eye over (4)
18 ... lights, aurora borealis (8)
20 Fault, puts through (anag.) (8)
21 Digital information (4)
22 Scottish loch (4)
23 Sixth planet (6)

Down
2 Level land (7)
3 Of the town (5)
4 Enterprise initiator (12)
5 Holiday maker (7)
6 Molten rock (5)
7 High cloud (12)
13 Seafloor flora and fauna (7)
15 Used to oxygenate water (7)
17 Small wood (5)
19 Regular behaviour (5)

Threshold concepts – difficult things

You may wonder why some topics appear in your degree programme, why some appear more than once, why you find some things easy and why other people don't. People who teach geography, earth and environmental science, and lots of other disciplines too, talk about the threshold concepts for their discipline, essentially things which are important to understand to master the discipline. This short chapter is a brief introduction to the concept because it might help you to understand a little bit more about how you learn, and how you learn about knowledge which you find difficult.

A threshold concept can be defined as 'something you need to understand in order to understand something else'. At its simplest, reading is a threshold concept for education, understanding that numbers and symbols kick-starts understanding in maths. Think of passing a threshold as a 'light bulb' or 'Eureka' moment in your life.

Tutors think of threshold concepts as things which students find difficult to understand, perhaps because it's counterintuitive, or it's a completely new idea, or it's just complicated. A tutor knows that if a student can understand this concept it will transform their understanding of that particular part of the discipline. This is called transformative knowledge.

How do you know when you've passed a threshold? Probably because you feel comfortable talking about the ideas in more academic or technical language. It is the difference between learning Spanish, and thinking and speaking in Spanish. Another advantage of passing the threshold is it can help you to see how different elements of a module or programme link together. It is about understanding in more depth and detail.

What is important to understand is that you rarely move from not understanding to understanding in one step. Research shows that transforming your understanding, or crossing a threshold, follows on from working with lots of pieces of information in different ways.

> It made sense when I listened in the lecture, and now I don't understand anything.

"In short, there is no simple passage in learning from 'easy' to 'difficult'; mastery of a threshold concept often involves messy journeys back, forth and across conceptual terrain." (Cousin 2006: 5)

Cousin points out that the reason why lecturers seem to be so involved with their material is that they have passed a series of thresholds in their understanding of their subject. And that makes it hard for them to understand and remember what it's like before things begin to 'click into place'.

University gives you the time to let lots of things click into place, transforming your understanding and through research you add to the 'conversations' between people that develop new ideas and understandings. You will probably transform the way that you approach learning while at university. You will do most learning yourself, in your own way, following reading and research suggestions made by lecturers.

This chapter has some brief examples of GEES students' thoughts on what is difficult, which may help you in thinking about how you approach university research.

3.1 Scale issues

An orrery is a physical model of the solar system planets, which helps people to understand how the planets move around the sun, as are plan diagrams (Figure 3:1). But these models are confusing because they mess with the scale. If the sun is modelled as a 20 cm diameter ball, then the Earth is the size of a peppercorn, Mars and Mercury are pinheads, and the distances between the sun and the planets is so

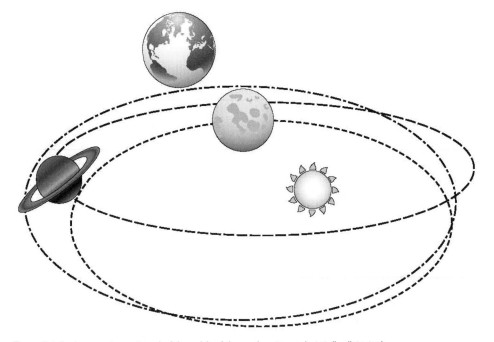

Figure 3:1 An inaccurate portrayal of the orbit of three planets, scale totally distorted

big that you couldn't see the peppercorn Earth while standing by the model sun. An orrery makes the planets look bigger than they are and very much closer together than they are. This makes space travel seem much more possible than it is. (See Ottewell 1989, and flickr: orrery video.)

> **Deep time:** Geological time, age of the planet, numbers are huge and hard to imagine.
>
> **Relative time:** Subdivisions of geological time defined by the relative position of rocks, and fossil evidence.
>
> **Absolute time:** Geological ages measured as accurately as currently possible, e.g. through radiometric dating of rocks and fossils.

The diameter of the earth and the depth of the atmosphere were very hard to understand before the first photos of the earth were taken from the moon. These showed how fragile and shallow the atmosphere is with respect to the diameter of the earth. (Google or MSN: earth atmosphere from space pictures.)

The Mohs scale of hardness (Chapter 28:8), used by geologists to compare rock mineral hardness, is relative. This is easier to understand once you hold and compare the different rocks.

The problem is that you cannot stand outside the solar system and see it *all*, you need to use your imagination!

Understanding space is fundamental in all the GEES subjects. Understanding relationships in 3D is recognised as difficult for many students, which really matters when constructing cross-sections and 3D plots. In geological map work, sedimentology, palaeontology and other modules, 3D practical work is quite common so that people get lots of practice and become familiar with two- and three-dimensional graphics.

Hydrologists need to understand how water flows three dimensionally through rocks and soil to understand groundwater and run off processes. Geomorphologists need 3D understanding to analyse landslide and glacier movements.

> Remember the Geological Time Periods:
>
> **P**ink **C**amels **O**ften **S**it **D**own **C**arefully, **P**erhaps **T**heir **J**oints **C**reak **P**ainfully, **E**arly **O**iling **M**ight **P**ossibly **P**revent **R**heumatism
>
> Precambrian Cambrian Ordovician Silurian Devonian Carboniferous Permian Triassic Jurassic Cretaceous Paleocene Eocene Oligocene Miocene Pliocene Pleistocene Recent

The length of geological time, the time taken for plates to move, for mountains to build and evolution to take place is difficult to imagine. The Holocene is a tiny fraction of geological history. Earth scientists talk about deep time, absolute time and relative time when constructing and reconstructing Earth history (Dodick and Orion 2006).

Take some time to think about how you imagine time and space.

3.2 Fieldwork can be transformative

GEES degrees are characterized by fieldwork. The aim is to show people what is really there. You escape the abstract notions formed through reading, and 2D lecture theatre slides become 3D. Understandings of geography, earth and environmental science matters happen in the field, when you get totally involved with the landscape and people. If you are not used to walking in the hills, scrambling through rivers, camping on glaciers and interviewing people, fieldwork can be difficult. Give yourself time to 'see' what is happening. Taking time to look, make notes, sketch (Chapter 25) and talk about what you see is important. Explore the area around your university, and the surrounding countryside by bus or rail. Urban or rural, they are all fantastic GEES landscapes, well worth a 'GEES-minded' look.

Every walk is like a field trip, even playing golf ... you just see things all the time. Now I want to know what is going on.

I didn't know what the glaciology stuff was about until we went to Svalbard. Then it all made sense – you can see it all happening ... and why they made us do a kit inspection first. Really cold.

Fieldwork tends to happen in separate weeks and separate modules. If cost was no problem most GEES material would be taught in the field, and integrated into most modules. Sadly this is not possible. Planning a holiday – think about where would help you understand a new landscape. Got a free weekend – explore the neighbourhood.

The field sketch session makes you look. I hated it because I didn't know what to draw and there is so much. It was when I looked at the tutor's sketch I began to understand ... He said 'it's not art ... more notes, and never mind the art'.

3.3 What university lecturers do

Feedback points out good points and mistakes. Making mistakes is a normal part of learning. Learning from mistakes is important. It is rare to get something right instantly ...

Remember that university lecturers are experts at exploring what is *not* known about their discipline (the essence of research). One of the functions of a university is to explore complicated matters and tease out the many dimensions and 'truths' involved, through the research process. Most researchers spend their careers developing ideas and exploring their validity (research), seeing ideas evolve as new information emerges. Lecturers are trying to sort out the confused understandings that exist in their subjects. The challenge for lecturers is to give students, usually accustomed to very structured school learning, insights into what is and what is not known, what is contested and what is agreed, and the confidence to work in the unexplored areas of the disciplines. It's brilliant when a little corner of science, social science or the humanities is made clearer.

3.4 People have different thresholds

Here are some reflections from GEES students either working in focus groups or from a post-university survey. Take some thinking time and as you read through. What are your thoughts? How might you follow up on your reflections? Keep your planning notebook handy as you read.

I suppose you could say my approach to uni involved having a great time playing hockey and then cramming at the last minute. So for me the threshold was revision, when I started doing the work. At that point you can see why different bits of the course got done, and it all began to make a bit of sense, but I don't think I really put everything together until about three months after the exams each year. Which was too late for good marks, but it did remind me why I'd wanted to do geography at uni.

… looking back you could say that writing an essay was a threshold for me. I did all maths and sciences at school and we didn't write an essay after I was about 13. It seemed like a junior school thing. But the lecturers just assumed I knew how to do it, so my marks weren't terribly good in first year.

… the people I lived with in halls in the first year weren't really interested in who I was. The girl who lived opposite was really brilliant at statistics and I didn't get it at all. So by Christmas I was ready to leave and thinking I couldn't pass anything. The trouble is, half the modules you do have numbers in them somewhere. It wasn't like that at school. Then you get on the field-class and do data, and you've got to do more stats. At least in a group there was somebody who could do them so as a group we all managed to get by. I think by the time we did the second-year field class I was understanding what was going on, but don't suppose I actually was enjoying it and it felt really unfair that other people could do it really easily.

… the first year skills class had that e-portfolio. Some people wrote loads and I couldn't work out what I was supposed to be writing. At the start it just didn't make sense that it would be useful to write down what we were doing. It really made sense of this last year doing the project, and for the work

placement. Then it was just totally brilliant. What really changed was I got used to writing about myself, and happy about making comments about what I was doing. What was really strange was doing some writing that no one was going to mark. Why bother? And it wasn't cool. Now it makes sense but then it just seemed rubbish.

I think I had a real issue with understanding things I couldn't see. You just can't see radioactivity, and dating fossils and rocks and stuff. That tutorial on gravity was just the … Look, gravity, it's just there, don't mess with it. I need to be told why I need to know what gravity is, because I'm happy with it the way it is.

If you have never heard of stereonets, how can you be expected to understand them in one practical?

I realized I needed to use a dictionary to get the words sorted out.

I just didn't understand how depressions work, until I was doing the school work experience and had to explain it to my class. Now there's all these 14-year-olds who understand depressions which is really cool, and so do I.

Numbers are OK, but what are equations? It was the second year lab guy reading them out as words in the practicals where it all made sense. No one really told me that the squiggly things in the equations all mean something. I guess that sounds really stupid but that's how it was.

We had all our first year in big lecture theatres and no one talked to anyone. In the second year we were working in groups all the time, and had to talk to each other. It's really hard working out that the teacher isn't going to tell you the answer, when obviously he knows what it is. You're used to that at school. If you wait long enough someone will tell you the right answer. Here it is all about working it out together, which was awesome when you got it.

3.5 Getting unstuck

> If you think you understood that first time round, you probably need to look at it again!

If you are not careful, you think about threshold concepts as something to get over in one step. All the evidence suggests that people learn in cycles, so you might like to think about how you get yourself unstuck. Savin-Baden (2008) describes working to get across threshold concepts as 'liquid learning'. Think of yourself as being flexible in order to deal with the real complexity of GEES materials. The point about university is to look at complicated matters. If you think explanations are simple then you may have missed a higher level of complexity. You may think about knowledge or information as blocks or units, but for most issues there are different understandings and interpretations of those blocks of knowledge. Statistics are taught in the first year so that they can be used in all years. Some topics come up in

> *It took me ages to realize that being confused at uni is normal. There are conflicting ideas about almost everything.*

a number of modules because lecturers appreciate that they are difficult, and they matter in different ways in different modules.

Different researchers have different opinions, research evidence and draw different conclusions. There are likely to be parallel understandings of most issues. Most GEES issues are influenced by understandings from other disciplines, for example chemistry, psychology, sociology and economics, which will provide further explanations and understandings of waste management, climate change, sustainability and transport systems. This diversity of understandings and ways of viewing GEES issues is referred to as contested knowledge, which is a threshold concept in its own right.

> *"Reflection is an important human activity in which people recapture their experience, think about it, mull it over and evaluate it. It is this working with experience that is important in learning..."* (Boud et al., 1985:19)

Learning at university is a threshold concept for most GEES students. The trick is to make time for mulling over, and giving yourself enough time for liquid learning so that you 'click' at the next level. Some topics need work until your brain 'gets it'. Give yourself the time to 'get it'.

3.6 References and further reading

Boud D, Keough R and Walker D 1985 *Reflection: Turning experience into learning*, Kogan Page, London

Cousin G 2006 An introduction to threshold concepts, *Planet*, 17, 4–5 www.gees. ac.uk/planet/p17/gc.pdf   Accessed 20 January 2011

Cousin G 2010 Neither teacher-centred nor student-centred: threshold concepts and research partnerships, *Journal of Learning Development in Higher Education*, 2, 1–9 www.aldinhe.ac.uk/ojs/index.php?journal=jldhe&page=article&op=view File&path[]=64&path[]=41   Accessed 14 January 2011

Dodick J and Orion N 2006 Building an understanding of geological time, in Manduca CA and Mogk DW (Eds) 2006 Earth and Mind: How geologists think and learn about the earth, *The Geological Society of America*, Special Paper 413, Boulder, Colorado 77–93

Savin-Baden M 2008 Liquid Learning and Troublesome Spaces: Journeys from the threshold? In Land R, Meyer RJ and Smith J (Eds.) *Threshold Concepts within the Disciplines*, Sense Publishers, Rotterdam

Words in Geo-words 1

Set 8 minutes on the timer (cooker, mobile ...) and see how many words you can make from:

SEDIMENT

Answers on p 296.

Maximizing free time

It isn't what you know that matters, it's what you think of in time.

This chapter involves lifesaving tips for people juggling uni, jobs, home and 'my life'. University is different from school and work. There is lots of time for free running, acting, being elected union secretary, playing the lute, extreme ironing and socializing, which is in part why many students find meeting coursework deadlines difficult. Time-management techniques are especially vital for people with major sports or social commitments, and/or part-time jobs. Developing your time-management skills should allow you to do all the boring tasks efficiently, like laundry and essays, leaving time for other activities. It is unlikely that any one idea will change your life overnight, but a few time-saving short cuts can relieve the pressure. Try something. Use your reflection and evaluation skills to identify what to do next and to assign time to research and read for uni.

The skills of project management are needed. Your life is your big project, as is an essay, module, field project and mapping exercise. Ideally one envisages the research/thinking/reading for an essay, project or dissertation moving linearly from inception to final report or presentation (Figure 4:1A). Regretfully the process is rarely this simple. The normal elements of life intervene, and the way you understand a topic changes as the research progresses. This makes the linear model (Figure 4:1A) totally unrealistic. The reality (Figure 4:1B) requires plenty of time for the research/thinking/reading process to evolve. Halfway through your research you may have to go back almost to the start, reconsider your approach and start a revised programme. Increasing your ability to manage your self and time, recognizing and adjusting to changing goalposts, are vital skills improved at university. Use **Try This 4.1**.

A The Optimist

| Idea | Plan | Research | Analyse | Interpret | Report |

B The Realist

Figure 4:1 The research process

⚙ TRY THIS 4.1 – Project manage your tasks

Find some Post-it notes, ideally in five different colours, and assign a different colour to: Tasks, People who can help, Milestone targets with dates, Finish date, Time for different tasks, Risks and barriers.

Now plan a holiday visit to … it's your choice. Put all the tasks that need doing on one colour (research, book, ask friends to come), another colour for milestone targets (visa obtained, booking made), and another for risks (no money, airline closes, cannot get time off work). The risks and barriers show where more planning (more task notes) or a plan adjustment (move things around) will be important. Put the Post-its on a large piece of paper/the wall/door … and move them around to get them into order. Involve other people, it is more fun and will give you additional ideas. The advantage of Post-it notes is that you can put the chart on your wall and move things around as you think of additional tasks or dates are missed (because this happens). And you feel good when completed activities are taken off.

Then use the same technique to plan your next essay or project.

See Figure 4:2 at the end of this chapter for two example charts.

4.1 Is there a spare minute?

Start by working out what time is available for research and study by filling out your timetable using **Try This 4.2**. Assume social and sport activities will fill every night and all weekend, and that arriving at university before 10.00 a.m. is impossible. The remaining time is available for research, reading, thinking, planning and writing, without touching the weekend or evenings. If you add a couple of evening sessions to the plan it will save money, due to temporary absence from bar or club, and help to get essays written. Divide this total free – oops, I mean *research* – time by the number of modules to get a rough target of the hours available for support work per module.

TRY THIS 4.2 – What spare/research time?							
Fill in your timetable: lectures, practicals ... the works. Block out an hour for lunch and a couple of 30-minute coffee breaks each day. Add up the free hours between 10 a.m. and 5 p.m. to find your Total Research Time.							
	Morning		Afternoon		Evening		
Monday							
Tuesday							
Wednesday							
Thursday							
Friday							
Saturday							
Sunday							

4.2 What do I do now?

Confused? You will be.

Diaries and timetables

University timetables can be complex, with classes in different places from week to week. A diary is essential. A weekly skeleton timetable will locate blocks of time for study (**Try This 4.2**). Use it to allocate longer free sessions for tasks that take more concentration, like writing, reading and preparing for a tutorial or workshop. Use shorter, one-hour sessions to do quick jobs, like tidying files, sorting lecture notes, summarizing the main points from a lecture, reading a paper photocopied for later, highlighting urgent reading, online searches, thinking through an issue and making a list of points that you need to be clearer about. Don't be tempted to timetable every hour. Leave time for catching up when plans have slipped.

Lists

Sort out what you need to do under four headings: Urgent Now, Urgent Next Week, Every Week, and Fun (see Table 4:1). If you tackle part of the non-urgent task list each week, you will be less overwhelmed by Urgent Now tasks at a later date. Have a go at **Try This 4.3**.

Urgent Now	Urgent Next Week	Every Week	Fun
Wednesday: Essay on Urban Poverty.	Read for tutorial on Nutrient cycling.	Tai Chi	Friday: cinema.
Friday: Palaeontology Report 3.	Find out about wind farms.	Ironing	Thursday: Dan's party, get card and beers.
		Supper	

Table 4:1 Keeping track of essentials

TRY THIS 4.3 – Essential or not?

Do a quick version of Table 4:1 for the next three weeks. Put a * against the items that you want to do in the next four days, and make a plan.

Create a diary template, online, on your phone, in your workbook with your regular commitments marked: lectures, tutorials, sport sessions, club and society meetings ... This provides the skeleton for planning. If weekly planning is too tedious, go for the 30-second breakfast-time, back-of-an-envelope version. It can really assist on chaotic days when classes of one hour are spaced out across the day. This can easily lead to the time in between disappearing. There are free hours but 'no time to do anything properly'. Completing short jobs will avoid breaking up days when there is more time. Try to set your day out something like this:

9	10	11 Lab.	12 Finish	2	3 Sort file	4 Africa	5–9
Lecture	Coffee Annie and Dan	Computer practical	computer practical Lunch	Tutorial	Read last week's Africa notes	seminar	Shop, night out

On days with fewer classes, two or more free hours gives you good research time for concentrated activities. A day might look like this:

10–12 Read and make notes for Marine Pollution essay	12 Lecture	1 Lunch email	2 Computer practical	3–5 Read and make notes for Marine Pollution essay	5–9 Wii, TV and phone calls

Knowing what you want to do in your research/thinking/reading time saves time. Deciding at breakfast to go to the library after a lecture should ensure you go to the right floor with your notes and reading lists. Otherwise you emerge from a lecture, take ten minutes to decide you would rather read about ecotourism than sustainability, discover you haven't got the ecotourism reading list, then look at the sustainability list to decide which library and floor to visit. All this takes 45 minutes and it is time for the next class.

4.3 Tracking deadlines

Deadlines are easily forgotten. For some people a term or semester chart that highlights deadlines that seem far away is helpful. Table 4:2 shows two chart styles. Which works for you? The first is essentially a list, the second one shows where pressure points build up. In this example, Weeks 10 and 11 already look full. The

Module	Assessment	Due Date	Interim Deadlines
EOE1202 Life Energy and Matter	Essay	10 Dec.	Check out a couple of background texts and case studies by 25 Nov. First draft by 11 Dec. Diagrams and revision 8 Dec. Final draft 8 Dec.
GEOG1070 Statistics	Practical	16 Nov.	Sort out the data set and run 1 Nov. Draft report 5 Nov. Final report 10 Nov.
GEOL1310 Planet Earth			

Week No/ Date	Social	Work-shops	Com-puting	Tutorial	Essays	Laboratory
(2) Oct. 6			Report 1 Friday			
(3) Oct. 13						Climate Report Wednesday
(10) Dec. 1	Rock Soc. Ball Thursday night	EOEE1040 worksheet Tuesday	Report 6 Friday	Presentation Aquaculture Tuesday		
(11) Dec. 8	End of Term Christmas shopping Hockey Club Dinner	GEOG1060 Test Tuesday	Report 7 Friday		National Parks essay Thursday	Sediment practical Thursday

Table 4:2 Example deadline planners for the semester

computer report due on Friday needs finishing before the Ball on Thursday! This second style highlights weeks where personal research/thinking/reading time is limited by other commitments. Start planning backwards. Look at Table 4:2. How would you adjust the tasks and timing to cope with these deadlines and the worst flu ever and the loss of your USB stick with all the notes? Add and move some Post-it notes.

4.4 What next?

The sooner you fall behind the more time you'll have to catch up.

Keep reviewing the plan. Look at the relative importance of different activities, so that you don't miss deadlines. Have a go at **Try This 4.4** as practice in prioritizing. Reflect on where you could re-jig things to release two lots of 20 minutes. Twenty minutes may not seem much, but using this time to sort notes, reread last week's lecture notes or skim an article, is 20 minutes more than you would have done.

TRY THIS 4.4 – Priorities?

Using yesterday as the example, jot down the time you devoted to each task, amending the list to suit your activities, and what priority they should have had. 1 is high priority and 5 is low priority. Then check to see how your priorities match the time taken.

Life	Hours	Priority	GEES Degree	Hours	Priority
Cooking and eating			Talking about ideas		
Sleeping			Reading		
Shower/Dressing			Browsing in the library		
Exercise			Thinking		
Travelling			Sorting lecture notes		
TV			Writing		
Reading for fun			Lecture attendance		
Cleaning the flat			Computer practicals		
Washing/Ironing			Laboratory practicals		
Phone calls			Planning time		
Facebook			Online searches		

Tackling **Try This 4.4** might encourage you to use a day planner, see Table 4:3 below. Write a 'to do' list for tomorrow, then prioritize your activities. What is important? What should be done first? Can you double task, reading while the washing tumbles around? Put some times against the activities. Ticking off completed jobs feels good!

Priority	MUST DO!	When
1	Post Mother's Day card	On way to college
2 =	Read Chaps 3–5 of Wave Power by Sue Nami	10–12
2 =	Launderette visit	10-12
3	Look at ecology notes for seminar	12–1
4	Check email	After lunch
1	Lecture 2.00 in Main Lecture Theatre	2.00
5	Spell and grammar check tutorial essay	After supper
6	Sort out practical notes	After supper
7	Make a list of jobs for tomorrow	After supper

Table 4:3 An organizer like this?

If you can do a task immediately and easily, that is usually efficient. Just do it. Generally it helps to allocate larger tasks to longer time chunks and leave little tasks for days that are broken up. Do not procrastinate: 'I cannot write this essay till I have read ...' is a lousy excuse. No one can ever read all the literature on any GEES topic, so set a reading limit, write, and then go clubbing.

Do you do:

1 Hard tasks first?
2 Easy tasks first to warm up to tricky ones?

or do you:

3 Make a list and then pick tasks at random?

Do what suits you, there is no right way. Find out what works for you.

 TOP TIPS

☺ Get an alarm clock/buzzer watch/timetable/diary.

☺ Plan weekends and time off well ahead; sport, socializing and shopping are critical. Having worked hard all week, you need and deserve time off. Following a distracting, socially rich week, maybe there is time for some study. Sunday can be a good time to draft a report, and a great time for reading and thinking as very few people interrupt.

☺ Filing '... so many modules, so many handouts, my room looks like a recycling depot'. Take 10 minutes each week to sort out notes and papers in your room and on your computer.

☺ Sort out when and where you work best. When are your high and low periods of concentration? High concentration times are best for study. Low concentration times are great for washing dishes. When your flat is noisy, find a quieter spot or do easy tasks like planning.

Sorting out my notes each Sunday made a fantastic difference to year 2. I could find stuff.

☺ If your mates are chatting in the geology library, go to the biophysics library. Avoid interruptions in your high concentration time.

☺ What do you find difficult? Sort out these tasks, do them first in your high concentration time. Divide awkward tasks into manageable chunks and tackle each one separately. Finishing parts of a task ahead of time gives you more opportunity to think about the geographical or geological interpretations (better marks). Most tasks that seem difficult become more difficult because they are left until time is short, and time pressures make them more tricky to do well. Get them out of the way.

☺ Short study times are good. Break up your day into one- and two-hour blocks. User the timer on your mobile or cooker. Work hard at the reading and writing, then have a break.

☺ Double tasking – view 'dead time', when walking to university, at the laundry or cleaning the bathroom, as 'thinking opportunities'. Plan an essay, mentally review lecture ideas …

☺ Vacations. Recover from term. Have a really good holiday. Think about dissertation possibilities.

☺ Time has a habit of drifting away. Can you limit lost time when the pressure is on? Minimize walking across campus. Ask 'Is this a trip I need to make? Could I be more efficient?' Make an agreement with a friend to do something in a certain time and reward yourselves afterwards.

☺ Be realistic. Most days do not map out as planned, things (people) happen, but plans make you more effective most of the time.

☺ Plan to be spontaneous tomorrow!!

Try some of these ideas, give them a real go for three weeks. Then reread this chapter, consider what helped and what did not, and try something else. Find a routine that suits you and recognize that a routine adopted in your first year will evolve in the following months and years. A realistic study timetable has a balance of social and fitness activities. Don't be too ambitious. If there was no reading time last week, finding 30 minutes to read one article is a step forward.

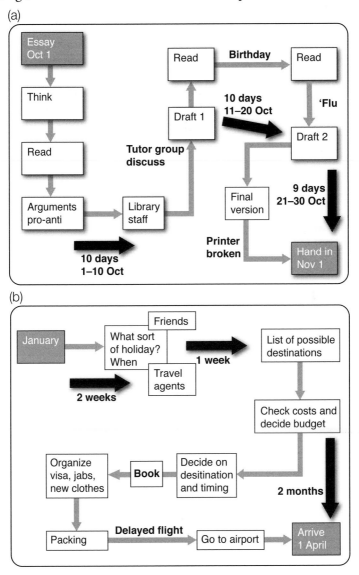

Figure 4:2 Project management plans for (a) completing an essay (b) booking a holiday

4.5 Further support

Kuther T 2010 Simple Steps to Master Your Use of Time, Aboiut.com: Graduate School, http://gradschool.about.com/cs/timemanagement/a/time.htm   Accessed 15 February 2011

LearnHigher 2010 Time Management – Resources for Students, http://learn higher.ac.uk/Students/Time-management.html   Accessed 15 February 2011

NUS 2010 Freshers and Settling In: Time Management for Students, http:// www.nus.org.uk/Student-Life/Freshers--Settling-In/Time-management-for-students/   Accessed 15 February 2011

UNSW 2010 Time Management, The Learning Centre, The University of New South Wales, http://www.lc.unsw.edu.au/onlib/time.html   Accessed 15 February 2011

Geolinks 1

By changing one letter at a time and keeping to real words, you move between the two terms. Answers on p 296.

B	A	L	I		U	I	S	T		B	U	T	E		E	I	R	E
M	U	C	K		L	O	N	G		F	A	R	N		M	U	L	L

5

Researching in libraries and online

And therefore education at the University mostly worked by the age-old method of putting a lot of young people in the vicinity of a lot of books and hoping that something would pass from one to the other, while the actual young people put themselves in the vicinity of inns and taverns for exactly the same reason.

(Pratchett, 1994, *Interesting Times*)

Once you discover that all the notes you made so conscientiously are completely unintelligible, or you did not quite make it to a lecture, getting to know the library might be a good idea. Inconveniently, GEES students find texts in many parts of the library. Geoscientists are likely to be heading to law, physics and engineering, geographers need books from agriculture, sociology, law, politics and civil engineering, and so do environmental scientists. This usually means books are in many locations and sometimes in different buildings. Then there are all the electronic resources. You can find e-books and access libraries internationally. Great fun – and potentially a good way to waste time. There is a maze of information, but finding the way around is not always obvious, but you CAN DO THIS. Skills you will draw on include researching, evaluation, information literacy, information retrieval, IT, flexible thinking and scheduling.

University libraries can seem scary and confusing. Most people feel very lost for the first few visits. This chapter gives information about library resources and research strategies, tips and hints that will, hopefully, reduce the mystery.

5.1 Library resources

Libraries are accessed in person and from your computer. For most library visits you need a library card to get in and out, cash or card for photocopying, pen, paper, USB stick or laptop for notes. Remember to watch your bags; the opportunist thief finds a library attractive as people leave bags while searching the shelves.

My first year was a real mess. It didn't seem cool to go to the library, it was really confusing. My tutor took us round when he realized we didn't know what to do … and then it was OK.

A few minutes with guided tours, watching videos and online explanations of your library's resources, and tips on accessing library and online documents will save you hours. Ask library staff to show you how to use the catalogue and search engines. Use Google Scholar.

Books and journals

At first I didn't see why you would go to the library when most of the reading was online ... I stayed in bed with the laptop to ... do stuff. But I don't get so much done at home, the library has so many more books ... I can get more done if I am there.

One of the changes from school to university is the emphasis on reading articles in journals. Journals contain collections of articles written by experts, published in every area of academic study. They are the way in which academics communicate their thoughts, ideas, theories and results. The considerable advantage of a journal over a book is that its publication time is usually six months to two years. Recent journals contain the most recent research results, which is really exciting. Check the location of:

✓ Recent issues of journals or periodicals. These may be stored in a different area of the library to the back copies. At the end of the year they are bound and join the rest of the collection. Reading recent issues can give a real feel for the subject and topics of current research interest.

✓ Government publications with endless tables of vital information for a geographer.

✓ Oversized books which do not fit on the shelves, are often filed as quartos at the end of a subject section. They are easy to miss.

✓ Stack collections containing less commonly used books and journals, usually somewhere else.

 TOP TIP

→ Reserve popular books – and read them!

Catalogues

Cataloguing systems vary between universities. Happily every library has handouts about how to retrieve material which involves searching via the online computer catalogue. This will show you where the book should be shelved and whether the copies are on loan or not. Before searching, highlight the papers and books on the reading list you want to read so search time is quick (see Chapter 6). If the books you want are out, check the shelf references for other texts you can substitute. The library catalogue is accessible through your campus computer network, letting you do bibliographic searches and mark up reading lists while the library is shut.

How do you know which items on a reading list are in journals and which are in books?

There is a convention in citing references, used in most texts and articles, that distinguishes journal articles from books, and from chapters in edited books. Traditionally, a book has its TITLE in italics (or underlined in handwritten text), a journal article has the title of the JOURNAL in italics, and where the article is a chapter in an edited book the BOOK TITLE is in italics:

Book:

Jackson AV and Hewitt CN 2009 *Atmospheric Science for Environmental Scientists*, Wiley-Blackwell, Chichester

Journal Article:

Bailey AJ 2009 Population Geography: Lifecourse Matters, *Progress in Human Geography*, 33, 3, 407–418

Belt ST, Massé G, Vare LL, Rowland SJ, Poulin M, Sicre M-A, Sampei M and Fortier L 2008 Distinctive ^{13}C isotopic signature distinguishes a novel sea ice biomarker in Arctic sediments and sediment traps, *Marine Chemistry*, 112, 158–167

Book Chapter:

Paul F and Hendriks J 2009 Detection and visualization of glacier area changes, in Pellikka P and Rees WG (Eds.) *Remote Sensing of Glaciers*, Taylor & Francis, London, 231–244

Parkin J 2010 Planning Walking Networks and Cycling Networks, in Givoni M and Banister D (Eds.) *Integrated Transport, From Policy to Practice*, Routledge, London 163–176

For example, to find the Paul and Henderiks paper, search for Pellikka and Rees (2009) *Remote Sensing of Glaciers*. Remember when searching to look up the *italicized* item first. You will never find the title of a journal article in a library main catalogue, but you will find the journal title and its library shelf location. Look for numbers, as here 74, 1, 149–155 indicates volume 74, issue 1, pages 149–155; books do not have this clue. Where there are no italics the game is more fun; you have to work out whether it is a journal or book you are chasing. (All students play this game; it's a university tradition.)

Unfortunately you cannot take out all the books at the beginning of term and keep them for the whole term. Find out what you can borrow and for how long, and what is available at other local libraries, the city or town library, for example.

Many libraries offer short-loan arrangements for material that lecturers have recommended as essential. Return books on time. Fines are serious, especially for restricted loans, and a real waste of money. When you need a book RECALL it. It encourages people, especially lecturers, to return them.

5.2 Electronic resources

Each university subscribes to a selection of e-resources and databases. Most are accessed with unique university passwords. You need to look at the resources online at your university library website and get used to the system. Remember that downloading papers, abstracts, reports, and articles is not a substitute for actually reading and making notes and learning the contents.

Materials available online are cited on reading lists with their URL and the date that they were accessed. This is important because web pages are updated and readers need to know which version you are referring to:

Keohane RO and Victor DG 2010 The Regime Complex for Climate Change. Discussion Paper 10–33. *Harvard Project on International Climate Agreements.* http://belfercenter.ksg.harvard.edu/publication/19880/regime_complex_for_ climate_change.html?breadcrumb=%2Fpublication%2Fby_type%2F discussion_paper   Accessed 15 February 2011

Many universities have online reading lists for modules that hotlink directly to the web source making searching very fast. Remember to look at other articles in the same journal to get a broader feel for what is available – the electronic equivalent of browsing the library shelves.

Online searches are made using the title, author or keywords. Before searching, make a list of keywords and decide if you need to search for English and American spellings. Searching for 'mountain bikes erosion' yielded 92,400 hits. Boolean operators speed up and refine a search, cutting out irrelevant sites (Figure 5:1). Entering 'Mountain+bikes+erosion' in Google Scholar will find 10,600 articles and patents. Refining this to articles published since 2006 cuts the list to 3740. Still too many so add further terms to focus your research list. Access to online resources which are so diverse means that each student finds a different combination of references and case studies.

There are three main Boolean operators used to refine searches:: + , – or **AND OR NOT**. Using Migration **OR** transmigration **AND** gender **NOT** Latin America should locate material on migration and gender issues from areas other than Latin America. Use **OR** when there are synonyms, and **NOT** to exclude topics. Using root words, like environ*, will find all the words that have environ as the first seven letters: environment, environmental and environs. Beware using root* too

Figure 5.1 Refining searches with Boolean operators

liberally. Poli* will get politics, policy, politician, political which you might want, and policeman, polite, polish and Polish, which you might not. Think about common synonyms 'cycle, bike, bicycle, tandem' unless when you use cycle you are wanting 'cycle, series, sequence, rotation'. It is worth an extra few minutes thinking about other terms to include in a search.

American and English spellings can be a nightmare; use both in keyword searches. Here is a starter table, but add to these as you find them:

English	American	English	American	English	American
artefact	artifact	defence	defense	metre	meter
behaviour	behavior	dialogue	dialog	mould	mold
catalogue	catalog	draught	draft	plough	plow
centre	center	enclose	inclose	sulphur	sulfer
cheque	check	enquire	inquire	traveller	traveler
colour	color	foetus	fetus	tyre	tire
counsellor	counselor	labour	labor	woollen	woolen

Finally, there is the geographical problem of changing place name: Peking and Beijing, Ceylon and Sri Lanka, Burma and Myanmar, Canton and Guangzhou. City names, especially those in the former Soviet Union, require careful cross-checking through atlases, with map curators or *The Statesman's Year-Book* (Turner 2011). There are many European examples that can cause confusion, for example Lisbon and Lisboa, Cologne and Koln, Florence and Firenze.

The web has proved to be the most time-wasting but fun element of a degree, while giving the comforting feeling of being busy on the computer all day. Book-marking 'favourite' pages will save you having to search from scratch for pages you use regularly. You should be able to email documents to your own file space. If you open web pages and word-processing packages simultaneously, and cut and paste between the two you can save printing costs, BUT beware of plagiarism, see p 128. This copy is in the original author's words – not your summary (see Chapter 8). It must be properly referenced and shown clearly in "..." (quotation marks).

TRY THIS 5.1 – WWW resources for GEES students

Explore some of these sites. If a site address is defunct, use a search engine and the site title to locate the updated address, e.g. British+Geological+Survey. (All sites accessed 15 February 2011)

BBC news online at http://www.bbc.co.uk/news

British Geological Survey at http://www.bgs.ac.uk/

Eldis – development policy, practice and research at http://www.eldis.org/

Environmental information for geotechnical, environmental, hydrogeology, geology, mining and petroleum topics at http://www.geoindex.com

Latin America information at http://www1.lanic.utexas.edu/

New Scientist at http://www.newscientist.com

Planet Earth online at http://planetearth.nerc.ac.uk/

Scottish Wetlands Archaeological Database http://xweb.geos.ed.ac.uk/~ajn/swad/introduction.html

Soilscapes at http://www.landis.org.uk/soilscapes/

The UK Meteorological Office at http://www.metoffice.gov.uk/

The United Nations Department of Economic and Social Affairs, Statistics Division at http://unstats.un.org/unsd/default.htm

US Census Bureau at http://www.census.gov/

WARNING!

☠ Accessing databases can be totally useful or utterly frustrating. Most are accessed with an authorized login and password. Do not get frustrated: libraries and departments cannot possibly afford to pay for access to all the sites. Not being able to access a specific item will not cause you to fail your degree.

☠ The fact that a database exists is no guarantee that it holds the information you need.

- Some www documents are of limited quality, full of sloppy thinking and short of valid evidence, some are fine: be critical. Anyone can set up a www site. Look for reputable sites, especially if you intend to quote statistics and rely heavily on site information. Government and academic sites should be OK.
- Think about data decay – which period of time does the information relate to? If, for example, you have economic or social data from the 1980s for Bosnia, or the old USSR, it will be fine for a study of that period in those regions, but of negligible value for a current status report. Check dates.
- Do not plagiarize. You **can** cut and paste from the Internet to notes and essays, but **if** you cut and paste, the source(s) must properly acknowledged. See pp 76 and 157.

5.3 Research strategies

There are oodles of background research documents for just about every geograph-ical topic, usually far too many. The trick in the library is to be efficient in sorting and evaluating what is available, relevant, timely and interesting.

A library search strategy is outlined in Figure 5:2. Look at it carefully, especially the recommendations about balancing time between searching and reading. Library work is iterative. Remember that online searches can be done when the library is closed but your computer is active. Become familiar with your local system; use **Try This 5.2** and your own university library site; or start with the links to University of Leeds (2011) or University of York (2011). Good library research skills include:

✓ Using exploration and retrieval tools efficiently.
✓ Reading and making notes.
✓ Evaluating the literature as you progress.
✓ Recording references and search citations systematically, so that referencing or continuing the search at another time is straightforward.

> **TRY THIS 5.2 – Library search**
> Choose any topic from one of your modules. Make a list of three authors and six keywords. Search for the papers and books in your library. Is there an interesting paper which is not on the reading list? Set a maximum of 30 minutes for online searching, then read and make notes for at least two hours.

It is possible to spend all day searching online. You will acquire searching skills, discover there is a paper with the ideal title in a library in Australia, or a foreign language, and have nothing for an essay. Ignore enticing www sites initially. For most undergraduate essays and projects, the resources in the library (paper or online) are more than adequate; read these articles first. Look at wider resources later, after the first draft is written. You cannot access and read everything: essays

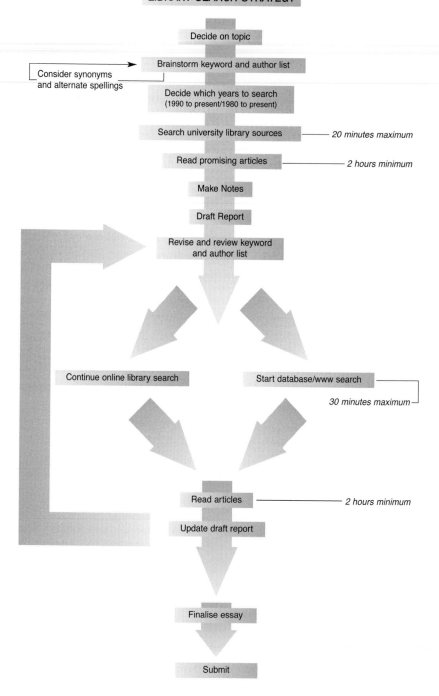

Figure 5:2 Library search strategy

have short deadlines and reading time is limited. The trick is to find documents available locally, with reliable origins and at no cost.

Wikipedia

Wikipedia can be a good starting point but BEWARE. There are some very good sites. Many of the more academic sites are looked after by university lecturers keen to improve understanding of complex issues. There are some sites where groups of students provide updates. BUT there are many sites where the information is unreliable and biased, where you are reading someone's personal opinion or advertising spin rather than an evidenced argument. University research should always be supported by academically respectable materials (add *references*). Use Google Scholar.

 TOP TIP

→ Do not quote from Wikipedia unless you are desperate. Use it as a starting point. Cite the linked articles if they have real academic relevance and authority.

Reading lists

Module reading lists are long, with lots of alternatives. This is vital where members of large classes want to access documents simultaneously. It gives you choice, and each student consults a different combination of texts. Some lecturers give quite short reading lists; these may include essential reading items, and a list of authors and keywords. This approach encourages students to explore the available literature independently. Where the reading lists are of the first type, it is wise to view it as a big version of the second!

Check whether the lecturer has provided references from the last two years. Most good lecturers do not do so because they want to see which students have the initiative and interest to follow up on topics, to locate new research and cite it in essays, reports and exams. This allows them to credit (*give more marks to*) people who are developing as effective researchers.

Review articles

Review articles, especially those in *Progress in Physical Geography*, *Progress in Human Geography Science of the Total Environment* and *Earth Science Reviews* give illuminating syntheses of recent literature, and point you to other references. Some journals have themed issues, or review article sections, e.g. *Environmental Science & Policy* and *Environment and Planning A*. You may start out with one reference, but in a themed volume there will be papers on interrelated topics that are, at the least, worth browsing.

Linked reading

If you happen to read Greenough *et al.* (2010) you may quickly realize that you need to read Greenough's earlier papers, and those of other authors they cited, to get a clearer idea of the context and science. It may seem obvious, but having read article A you may need to read article B to understand A. Cross-referencing reading is a normal learning process. Remember to take accurate notes of the references that you follow up; they may prove to be the ones you cite in your writing.

 TOP TIPS

→ Be particularly critical of sources that may have a bias or spin.

→ Keep track of your reading, and a balance between searching (20 minutes) to real reading and note-making (2 hours).

5.4 Why am I searching?

Library searches are never done in isolation. Before starting, review your reasons for searching and focus your keyword search. Put a limit on your searching time and on the types of document to include. Suggestions on this front include:

✓ Module essays: start with the reading list, and only explore further when you have an initial draft. Look critically at the gaps in your support material and use **Try This 5.3**.

✓ Reading 'state-of-the-art' studies exploring the current state of knowledge on a topic; limit searches to the last 2–5 years.

✓ Reading an historical investigation of the development of an idea, considering how knowledge has changed over 10, 20 or more years; aim for a balance between the older and newer references.

✓ Reading a literature review should give the reader an outline of the 'state' of the topic. It may have a brief historical element, mapping the development of the subject knowledge, leading into a more detailed resume of research from the past 5–10 years.

TRY THIS 5.3 – Self-assessment of a library search

Keep a tally of sources and authorities when doing a library search or preparing a report or essay. Check for an advantageous balance of recent citations and that all the appropriate sources are used, in addition to those on the reading list.

Sources	Books	Journal Articles	Websites	Reports	Other
Articles on reading list					
Articles not on reading list					
Vintage of sources	1980–89	1990–99	2000–05	2006–10	2011–date

Do not forget international dimensions: the topic might be theoretical (urban development, social housing, migration, urban climatology), but there may be regional examples that are worth considering, so check out the journals of other countries. *The Singapore Journal of Tropical Geography, Australian Journal of Earth Sciences, The Canadian Geographer, Canadian Journal of Earth Sciences, Canadian Journal of Forest Research, New Zealand Journal of Ecology, Annals of the Association of American Geography* and *Irish Geography* are all international journals, which contain articles reflecting national and regional concerns.

WRITE UP AS YOU GO, keep noting and drafting, and keep a record of references in full. Remember to add the reference, dates and pages on notes and photocopies.

5.5 References

Greenough JD, Fryer BJ and Mallory-Greenough L 2010 Trace element geochemistry of Nova Scotia (Canada) maple syrup, *Canadian Journal of Earth Science*, 47, 8, 1093–110

Turner B (Ed.) 2010 *The Statesman's Year-Book 2011*, 145th Edn., Palgrave Macmillan, London [This may be available at www.statesmansyearbook.com/public/ if your library has a site licence]

University of Leeds 2011 Web Searching Tutorial, Skills library, at http://library.leeds.ac.uk/documents/tutorials/web_search/web_searchindex.html   Accessed 15 March 2011

University of York 2011 Research methods and resources at http://www.york.ac.uk/library/subjectresources/researchmethods/   Accessed 15 March 2011

Geo-codeword 1

Replace the numbers with letters starting with the three indicated below. Complete the grid below to find GEES-related terms. Answers on p 296.

1 L	2	3	4	5 D	6	7	8	9	10	11	12	13
14	15 O	16	17	18	19	20	21	22	23	24	25	26

17	5	13	7	4	11	■	17	25	17	11	17	20
8	■	■	24	■	14	■	7	■	10	■	3
20	3	17	1	10	11	23	■	1	15	5	14	3
17	■	11	■	1	■	2	■	11	■	17	■	22
4	26	15	17	1	■	4	18	7	17	1	1	■
10	■	1	■	■	23	■	20	■	■	19
15	15	1	10	11	3	■	5	3	9	11	3	20
16	■	■	20	■	17	■	■	20	■	10
■	10	4	1	17	16	5	■	5	20	17	22	11
6	■	12	■	2	■	7	■	15	■	10	■	3
10	21	17	14	3	■	1	17	21	10	16	17	20
16	■	21	■	6	■	11	■	21	■	■	10
19	15	2	2	3	20	■	21	23	15	2	10	17

6

Effective reading

I read about the hazards of typhoons. I was so frightened,
I gave up reading.

Everyone reads; the knack is to read and learn at the same time. Take a minute to think about where and how you read, and consider how effectively you learn as you read. You can develop your ways of reading. There is a vast amount of information to grapple with in all the GEES degrees. Reading, thinking and note-making are interlinked activities. There are suggestions here that will help you to pick and choose reading options. Remember to slow down and enjoy reading. Give your brain the space and time to absorb, understand and interact with the information. This chapter concentrates on techniques that will help you to learn as you read.

> TOP TIP
>
> → Carry reading lists at all times.

6.1 Reading lists

Inconveniently most reading lists are alphabetical, but you need to sort out what to read urgently, and where to find the materials: online, in the library and the library location. Target more items to read than you can reasonably do in a week, so that if a book is unavailable there are alternatives.

> *There were book sales in the department at the start of the year. The second year students were really helpful, told us what was worth buying.*

Reading lists are often dauntingly long, but you are not, usually, expected to read everything. Long lists give you choice, which is especially impor-

> *I bought one book in first year. The three of us in our flat made sure that between us we got the main books from the library, and we shared them.*

tant where class sizes are large. Serendipity cheers the brain. If a book is out on loan, don't give up. There are probably three equally good texts on the same topic within the same library class number. Books and journal articles take time to go through the publication process. Do you have the latest text or edition? Look along the shelves. If you find a topic confusing reading another author's views and ideas can be very helpful. Remember that reading something is more helpful than reading nothing. RECALL essential texts.

Even where there is a recommended textbook, it will rarely be followed in detail. Reading a recommended book is a good idea, but watch out for those points where a lecturer disagrees with the text. Perhaps the author got it wrong, or our understanding of a topic has moved forward, or ideas have changed.

Activity	
	Tasks and Issues
Sort out reading lists	Prioritize modules. Find reading lists. Use highlighter/underline MUST READ and MIGHT READ items.
Plan reading time	Decide on time and places to read – library, online, armchair. Put times in your planner for the next three weeks.
Decide what to do first	Make today's plan.
Update task list	End of session/day review. Add in new papers to check to your list. Cross off those where you are happy with your notes

 TOP TIP

→ **READ FOR ALL MODULES!**

6.2 Reading techniques

Photocopying is no substitute for reading – but it feels really, really good.

There is a mega temptation to sit down in a comfy chair with a coffee and to start reading a book at page 1. THIS IS A VERY BAD IDEA. By page 4 you will have cleaned the cat litter tray, done a house full of washing, mended a motor bike, fallen asleep or all four and more. This is great for the state of the flat, but a learning disaster.

Everyone uses a range of reading techniques: speed-reading of novels, skip-reading headlines; the style depends on your purpose. As you read this section reflect on where you use each technique already. Effective study needs 'deep study reading'.

Deep study reading

Deep study reading (reading when you learn as you go) is vital when you want to make connections, understand meanings, consider implications and evaluate arguments. Reading deeply needs a strategic approach and time to think. Rowntree (1988) describes SQ3R which promotes deeper, more thoughtful reading. It is summarized in **Try This 6.1**. SQ3R is an acronym for Survey – Question – Read – Recall – Review. Give **Try This 6.1** a go; it will seem long-winded at first, but it is worth pursuing because it links thinking with reading in a flexible manner. It stops you rushing into unproductive note-making. You can use SQ3R with books and articles, and for summarizing notes during revision. You are likely to recall more by using a questioning and 'mental discussion' approach to reading. Having thought about SQ3R with books, use **Try This 6.2**.

Expect your reading rate to be much slower than normal. Tackling a few pages, understanding and remembering the ideas will be more useful than covering more pages but recalling less.

Browsing

Browsing is an important research activity, used to search for information that is related and tangential to widen your knowledge. It involves giving a broader context or view of the subject, which in turn provides you with a stronger base to add to with directed or specific reading. Browsing might involve checking out popular science, social science, and introductory texts. Good sources of general and topical geographical and environmental science information include *The Economist*, *New Scientist*, *New Internationalist* and the country and investment focus supplements in the *Guardian* and *Financial Times*. Browsing enables you to build up a sense of how the subject you are studying as a whole, or particular parts of the subject, fit together. Becoming immersed in the language and experience of the topic encourages you to think like a professional earth scientist or geographer.

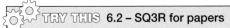 **TRY THIS 6 .1 – SQ3R**

SQ3R is a template for reading and thinking. Try it on the next book you pick up.

Survey: Look at the whole book before you get into the detail. Start with the cover: is this a respected author? When was it written? Is it dated?

Read the contents and chapter headings and subheadings to get an idea of the whole book and to locate the sections that interest you. First and last paragraphs usually highlight arguments and key points.

Question: You will recall more if you know why you are reading, so ask yourself some questions. What do I know? What else I want/need to know? What is new in this reading? What can I add from this book? Where does this fit in this course, other modules? Is this a supporting/refuting/contradictory piece of information?

Having previewed the book and developed your reasons for reading, decide whether deep reading and note-making is required, or whether scanning and some additions to previous notes, will suffice.

Read: This is the stage to start reading, but not necessarily from page 1 Read the sections that are relevant for you and your present assignment. Read attentively but critically. The first time you read you will not understand it all.

On first reading: locate the main ideas. Get the general structure and subject content in your head. *Do not make notes this time*, the detail gets in the way.

On second reading: chase up the detailed bits that you need for essays. Highlight, or make notes of, all essential points.

Recall: Do you understand what you have read? Give yourself a break, and then have a think about what you remember, and what you understand. This process makes you an active, learning reader. Ask yourself questions like: How do I explain this idea in my own words? How do I recall the key points (without rereading the book)?

Review: Now go back to the text and check the accuracy of your recall! How much you have really absorbed? Are the headings and summaries first noted the right ones, do they need revising? Do new questions about the material arise now that you have gone through in detail? Have you missed anything important? Do you need more detail or examples?

Fill in gaps and correct errors in your notes. Ask where your views fit with those of the authors. Do you agree/disagree?

The last question is 'Am I happy to give this book back to the library?'

TRY THIS 6.2 – SQ3R for papers

Adapt SQ3R for reading a journal article, and use it on the next article you read.

Scanning

Scan when you want a specific item of information. Scan the contents page or index, letting your eyes rove around to spot key words and phrases. Chase up the references and then, carefully, read the points that are relevant for you.

Skimming

Skim-read to get a quick impression or general overview of a book or article. Look for 'signposts': chapter headings, subheadings, lists, figures; read first and last paragraphs/first and last sentences of a paragraph. Make a note of key words, phrases and points to summarize the main themes; but this is still not the same as detailed, deep reading.

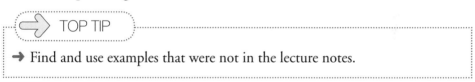

TOP TIP

→ Find and use examples that were not in the lecture notes.

Big questions

When reading, ask yourself:
- Is this making me think?
- Am I getting a better grasp of the material?
- Am I giving myself enough time to think and read?

If the answers are no, then maybe a different reading technique would help. Reading is about being selective, and it is an iterative activity.

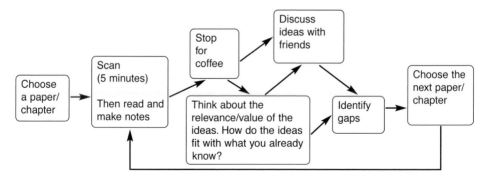

Figure 6.1: The reading plan

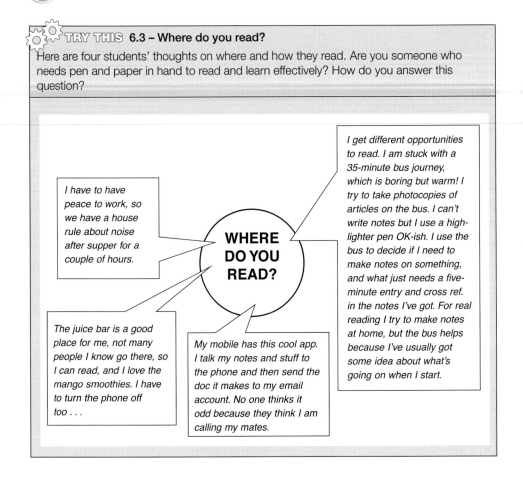

TRY THIS 6.3 – Where do you read?

Here are four students' thoughts on where and how they read. Are you someone who needs pen and paper in hand to read and learn effectively? How do you answer this question?

Building arguments

Academic journal articles and books are not thrillers. There should be a rational, logical argument, but rarely an exciting narrative. Usually, authors state their case and then explain the position, or argument, using careful reasoning. The writer is trying to persuade the reader (you) of the merit of the case in an unemotional and independent manner. You may well feel that the writer is completely wrong. You may disagree with the case presented. If so, do not 'bin the book'; make a list of your disagreements and build up your counter-arguments from other papers. If you agree with the author, list the supporting evidence and case examples. First class assignments report both sides of arguments (*with references*).

Most writers use cues or signposts to guide you to important points, phrases like: The background indicates ..., the results show ..., to summarize .... **Try This 6.4** will help you get comfortable with academic reading.

> ⚙ TRY THIS 6.4 – Cue Word Bingo
>
> Look through the book or journal article you are reading today, to find these cue words and phrases. How are they used? How many can you spot in a lecture?
>
> **Starter cue phrases, add more as you find them:**
>
> I shall outline the theory behind ... The next point is ... We must first examine and seek to quantify ... Consider the issues of ... There is little known about ... These include ... Recall ... Remember ... However ... This is the most significant advance ... We have shown that ... To summarize the principle points ... Restating the original argument in the light of this information we see ... We can conclude that ... We now know that ... The important conclusion was ...

6.3 How do you know what to read?

What do I know already?

Reading and note-making will be more focused if you first consider what you already know, and use this information to decide where reading can effectively fill the gaps. Use a flow or spider diagram (Figure 11.1, p.104 and Figure 17.1 p.169) to sort ideas. Put boxes around information you have already, circle areas which will benefit from more detail, check the reference list for documents to fill the gaps, and add them to the diagram. Then prioritize the circles and references, 1 to n, making sure you have an even spread of support material for the different issues. Coding and questioning encourages critical assessments and assists in 'what to do next' decisions.

> My friend, who does English, reads three books each week. How much should I be doing?

Be critical of the literature

Before starting, make a list of main ideas or theories. While searching, mark the ideas that are new to you with asterisks, tick those which reinforce lecture material, and highlight ideas to follow up in more detail. Ask:

- Is this idea up to date?
- Are there more recent ideas?
- Do the graphs make sense?
- Are the statistics right and appropriate?
- Did the writer have a particular perspective that led to a bias in writing?
- Why did the authors research this area?
- Does their methodology influence the results in a manner that might affect the interpretation?

Library, author and journal searches start the process (Chapter 5) and practice allows you to judge the relative value of different documents. After reading, look at the author and keyword list again. Do you need to change it? Exploring diverse sources will develop your research skills. Reading and quoting sources in addition to those on the reading list may seriously impress an examiner.

NARROW READING → predictable essays and reports → middling marks

WIDE READING → more creative, less predictable responses → higher marks (usually)

It does not usually matter what you read, or in what order. Read something.

How long to read for?

For most people, two hours is long enough to concentrate on one topic. A short article from *Nature*, *New Scientist* or *The Economist* should take less time, but some journal articles will take longer. With longer documents you need a reading strategy, and to take breaks. Use the breaks to reconsider the SQ elements of SQ3R and decide whether your reading plan needs amending. If you cannot get involved with a text then it is possibly because you cannot get to grips with the point of the writing, or do not know why you should be interested. STOP READING and skim the chapter headings, skim your notes, refresh your brain on WHY you are reading and what you want to get out of it.

6.4 Styles of writing

If you find an article difficult to read, it may be due to unfamiliarity with the topic, but also with the written style. Writing styles reflect the conversational language and approach adopted in sections of the disciplines. Styles range from very direct, typically scientific content with short sentences and information-rich content. Environmental law and ethics modules involve more legal writing. The more philosophical literature, in cultural geography for example, is much more discursive. Becoming familiar with the different languages of GEES specialisms is part of being a GEES student.

Whatever the style, look for the broad themes rather than the detail. The *Transactions of the Institute of British Geographers* serves the entire geography community and is worth browsing through to see the different styles of writing. *Science of the Total Environment* is always interesting and has contrasting writing styles. Earth science and geology journals have scientific styles. Think about the writing style as you read journal articles.

Don't be put off by a writer's style. Make sure you pick up the main message from each section. Be sure you are clear on the supporting and opposing arguments. Are things less black and white? Are there parallel arguments? Does the text give one side of the argument? Can you think of another side? Do you need to read something else to balance this author's view? If an article seems difficult, look at related, scene-setting materials and then reread the paper.

Think about how you read and take notes (see Chapter 7) from factual, short-sentence styles of writing, and from more philosophical or discursive styles. The trick is to adapt your reading and note-making style to maximize your learning. This needs practice. Sharing your notes with friends can be a useful way of seeing different approaches.

 TOP TIPS

→ Ask yourself: Do I understand this?

→ Ask at the end of a page, chapter, paper, tutorial, lecture … and not just at the end!

6.5 References and further reading

What do you call the bloke that stopped the river?

Buzan T 2006 *Speed Reading: Accelerate Your Speed and Understanding for Success*, BBC Books, Harlow, UK

Rowntree D 1988 *Learn How to Study: A guide for students of all ages*, 3rd Edn., Warner Books, London

Adam

Drop out 1

Remove one word from each column to make one word, and the remaining letters align to make six additional words across. Answer on p 296.

E	M	E	N	A	L	D
C	A	D	R	I	T	M
V	R	H	M	C	U	E
F	E	A	I	O	L	T
G	A	L	L	B	U	C
A	E	R	O	U	I	E
P	O	L	L	I	T	E

7

Making effective notes

I made a mental note, but I've forgotten where I put it.

Life is full of information whizzing around the net, videos, TV reports, specialist documentaries, lectures, tutorials, discussion groups and in books, journals and newspapers. BUT, just because an article is in an academic journal, in the library, or on a reading list, does not make it a 'note-worthy' event. Making notes takes time, if your brain is fully involved in asking questions and commenting on the ideas, you will start to learn

> **Note taking or note making?**
> **Note taking** happens when everything is new, you are trying to capture facts and ideas.
> **Note making** is when you look at your notes and make sense of them. You add, cross out, make links, and learn stuff.
> *When do you make notes?*
> **Where do you make notes?**

the material too. Noting is not just about getting the facts down, it involves identifying links between different pieces of information, contradictions and examples. Notes should record information in your own words, evaluate different points of view, and encourage the development of your own ideas and opinions. Note-making is a multipurpose activity; like snowboarding, it gets easier with practice. Good questions to ask when making notes include: 'Is this making me think?' or 'Am I getting a clearer understanding of the topic?' Notes are not usually assessed. The major exception is field notebooks which are normally assessed for all geology field classes, and which may be assessed on geography and environmental science courses.

Always record your sources when making notes.

Many people start reading and making notes without any sort of preview. A BAD IDEA. They make pages of notes from the opening section and few, if any, from later in the document. The first pages of a book usually set the scene. Notes are often only needed from conclusion and discussion sections. Sometimes detailed notes are required, but sometimes key words, definitions and brief summaries are enough. Use **Try This 7.1** to evaluate how the style and length of your note-making should change given different types of information, and consider how the SQ3R method (see p 60) fits in to your note-making process. Then look at **Try This 7.2** and reflect on what do you do already. What could you do in future?

TRY THIS 7.1 – Styles of note-making

What styles of notes are needed for these different types of information? There are three possible answers to kick-start your reflection.

Academic content – knowledge	Style of notes
Significant article but it repeats the content of the lecture.	None, it is in the lecture notes. BUT check your notes and diagrams are accurate. Did you note relevant sources, authors?
Fundamental background theory, partly covered in the lecture.	
An argument in favour of point x.	
An argument that contradicts the main point.	
An example from an odd situation where the general theory breaks down.	
A critically important case study.	
Just another case study.	
Interesting but off-the-point article.	A sentence at most! However, add a cross-reference in case it might be useful elsewhere.
An unexpected insight from a different angle.	
An example/argument you agree with.	
An argument you think is unsound.	Brief notes of the alternative lines of argument, references and case examples. Comment on why it does not work so the argument makes sense to you at revision.
A superficial consideration of a big topic.	
A very detailed insight into a problem.	

⚙ **TRY THIS 7.2 – How do you make notes?**

Look at this unordered jumble of note-making activities and ✓ those likely to assist learning, and put a ✘ against those likely to slow up learning. What do you do already? What could you use next time you write notes?

Leave wide margins	Code references to follow up
Identify what is not said	Store notes under washing
Compare and revise notes with study buddies	Annotate handouts
Do loads of photocopying	Write down everything said in lectures
Underline main points	Write shopping lists in lectures
Arrange to meet friends in the library	Ask questions
Make notes from current affairs programmes	Ignore handouts
Doodle	Jot down personal ideas
Make short notes of main points and headings	Use coloured pens for different points
Turn complex ideas into flow charts	Copy all PowerPoint slides
Ask lecturers about points that make no sense	Write illegibly
Highlight main points	Use cards for notes
Take notes from TV documentaries	Natter in lectures
Scribble extra questions in margins	Always note references in full
Revise notes within three days of lectures	Copy big chunks from books

7.1 Making notes from presentations

Field-classes, workshops, seminars and lectures are awash with presentation, and memory meltdown syndrome will loom. Make notes of important points; you cannot hope to note everything. Listen to case studies and identify complementary examples. Highlight references mentioned by the lecturer and keep a record of new words. Your primary goal in presentations should be to participate actively, thinking around the subject material, not to record a perfect transcript of the proceedings. Get the gist and essentials down in your own words.

Being critical involves questioning what you read; don't agree just because it is in print. Look for reasons why we should or should not accept the information as true.

Constructive criticism offers suggestions to further develop or improve a piece of research or writing.

Are the slides on the VLE? Brilliant. Reading through the slides in advance, and adding notes to them at a presentation will seriously help your learning. This is because your brain is already 'tuned in', not trying to work out what is going on. You start note-making at the lecture.

Lecture notes taken at speed, in the darkness of a lecture theatre, are often scrappy, illegible and usually have something missing. If you put these notes away at once, you probably won't make sense of them later. Plan time to summarize and clarify notes within a day of the lecture. This reinforces ideas in your memory, hopefully stimulates further thoughts, and suggests reading priorities.

7.2 Making notes from documents

Noting from documents is easier than from lectures, because there is time to think about the issues, identify links to other material and write legibly the first time. You can read awkward passages again, but you risk writing too much. Copying whole passages postpones the hard work of thinking through the material – wasting time and paper. Summarizing is a skill that develops with practice. Give **Try This** 7.3 a go next time you read a journal article; it won't work for all articles but is a start to structuring note-making.

TRY THIS 7.3 – Tackling journal articles

Use this as a guide when reading a journal article or chapter.

1 Write down the reference in full and the library location so you can find the journal again.
2 Summarize the contents in two sentences.
3 Summarize in one sentence the main conclusion.
4 What are the strong points of the article?
5 Is this an argument/case I can agree with?
6 How does this information fit with my current knowledge?
7 What else do I read to develop my understanding of this topic?

Think about where you will use your information. Scanning can save time if it avoids you making notes on an irrelevant article or one that repeats information you have already. In the latter case, a two-line note may be enough, e.g.

> *Withyoualltheway (2020) supports Originality's (2015) hypothesis with his results from a comparable study of the ecology of lemmings*

or

> *Dissenting arguments are presented by Dontlikeit (2020) and Notonyournellie (2020) who made independent, detailed analyses of groundwater data to define the extent of pollution from landfill leakage. Dontlikeit's main points are ...*

or

> *Wellcushioned (2020), studying 27 retail outlets in Somerset and West Virginia, showed price fixing to be widespread. His results contrast with those of Ididitmyway's (2018) report on prices in Bangkok, because ...*

In a report, essay or examination you have limited words and time. An essay with one case study as evidence is likely to do less well than one which covers a range of examples or cases, albeit more briefly, PROVIDED THEY ARE RELEVANT. Short notes help you to focus on arguments that are supported by a range of evidence (*references*). Make notes accordingly.

7.3 Fieldwork notes

All GEES students do fieldwork. Geologists probably do the most. *Accurate* fieldwork notes are vital academically, so that you can write excellent reports somewhere dry and warm with a table. A small hardback notebook is ideal, with a large clear plastic bag to keep it and your hand dry as you continue to write during monsoonal rain or a snowstorm.

Focus note-taking on recording your impressions and observations as accurately as possible. It's important not to miss things. For geologists it involves very accurate descriptions of the rocks, minerals, fossils and geological structures. Field sketches are really important. Your tutors will talk about creating a good field notebook in great detail, which may seem a little boring but professional geologists are making field observations all the time. Their notebooks are the legal record of their field observations. Getting this right is very, VERY important. Geology field classes prepare you for this professional workplace activity through the assessment of field notebooks. Make sure all pages are dated, the location of each site is clearly marked with map or global positioning system (GPS) coordinate, and sketches should have a compass direction to make it clear which aspect is involved.

Geographers and environmental scientists need to keep equally good records. It makes writing up reports from field activities easier. Stopping to sketch and make notes about what you are seeing encourages you to look in more depth at the landscape. Draw what is there, then add the human dimension as appropriate, traffic, types of industry, housing styles, ethnic and cultural information ... be aware of the landscape at different scales. You may be sketching one house, stream or hillside but make notes about the wider landscape, vegetation, housing and weather, too. It is not always obvious at the start what will be important for your final report. Make notes about what you see and what the lecturer is talking about.

There is a tendency to take photographs rather than make field sketches. Sketches (Chapter 25) are better because you can write notes on them, and you really look at the view. Photographs are helpful as backup but note why you took the photograph and what was important. It's very hard to see the detail and make an excellent sketch at a later stage from a photo. It's the activity of drawing and adding notes to the drawing that helps you to develop the GEES skills to understand a particular landscape.

7.4 Techniques

Note-making is an activity where everyone has his or her own style. Aim to keep things simple, or you will take more time remembering your system than learning about ammonites, arachnids or Argentina. Try some of the following (not all at the same time).

Which medium?

Hang on while I staple this note to my USB stick.

- Cards encourage you to condense material or use small writing. Shuffle and sort them for essays, presentations and revision.
- Loose-leaf paper lets you file pages at the relevant point and move pages around, which is especially useful when you find inter-module connections.
- Notebooks, field and lab notebooks keep everything together, but leave spaces to add new information, feedback, comments and make links. Index the pages so that you can find bits!
- Laptops and electronic organizers allow notes to be typed straight to file. This saves time later, especially with cutting and pasting references, but you may think less about the content …
- Mobile phones with voice recognition apps let you talk to the phone and upload to your email or files. Use these apps to dictate notes when you are on the move. It is fantastic for capturing ideas while running, less good in the middle of a hockey game.

Multicoloured highlighting

Colour-code to highlight different types of information on your handwritten notes, books and photocopies. Aim to distinguish key information and definitions, facts and figures worth learning, big ideas and links between things; information gaps – the things you want to find out more about. This approach requires four highlighter pens and consistency in their use, but it can be particularly useful when scanning your own documents. If this looks too complicated, use one highlighter pen – sparingly. On screen, changing the font colour or shading does the same task.

Mind maps

Turning notes into mind maps helps some people. Aim to create a visual representation of the ideas from your reading or lecture. Mind maps are especially useful for showing how the ideas are interrelated and the results are usually interesting, so that you think a little more. (Google or MSN 'mind maps' if this is a new idea for you.)

Coding

Coding notes assists in dissecting structure and picking out essential points. During revision, the act of classifying your notes stimulates thoughts about the types and relative importance of information. At its simplest, use a ** system in the margin:

****	Vital	**	Useful
?*	Possible	I	A good idea but not for this, cross-reference to ...

A more complex margin system distinguishes different types of information:

||	Main argument	B	Background or introduction
|	Secondary argument	S	Summary
E.G.	Case study	I	Irrelevant
[	Methodology, techniques	!!	Brilliant, must remember
R	Reservations, the 'Yes, but' thoughts		
?	Not sure about this, need to look at ... to check it out.		

Opinions

Note your own thoughts and opinions as you work. These are vital, BUT make sure you know which are notes from sources, feedback from the group, and which your own opinions and comments. You could use two pens, one for text notes and the other for personal comments. Ask yourself questions like: What does this mean? Is this conclusion fully justified? Do I agree with the inferences drawn? What has the researcher proved? What is s/he guessing? How do these results fit with what we knew before? What are the implications for where we go next?

Space

Leave spaces in notes, a wide margin or gaps, so there is room to add comments and opinions at another time. There is no time in lectures to pursue personal questions to a logical conclusion, but there is time when reviewing to refocus thoughts.

Abbreviations

Use abbreviations in notes but not essays. Txtspk may help here: intro. for introduction; omitting vowels: Glc^n for glaciation; $Hist^l$ for historical; or using symbols. You probably have a system already, but here are some suggestions:

+	And	=	is the same as
→	Leads to	xxx^n	xxxion as in precipitation
↑	Increase	xxx^g	xxxing as in pumping

↓	Decrease	//	Between
>	Greater than	Xpt	Except
<	Less than	←	Before
∴	Therefore	w/	With
?	Question	w/o	Without

Millions of unordered notes will take hours to create but will not necessarily promote learning. Aim for notes that:

✓ are clear, lively and limited in length.
✓ add knowledge and make connections to other material.
✓ include your own opinions and comments.
✓ are searching and questioning.
✓ guide and remind you what to do next.

Finally, feeling guilty because you haven't made some, or any, notes is a waste of time and energy.

TOP TIPS

→ **New words** – almost every GEES module has its specialist vocabulary. Keep a record of new words and check the spelling! The trick is to practise using the 'jargon of the subject' or 'geo-speak'. Get familiar with the use of words like ethnocentric, hysteresis, and pyroclastic. If you are happy with geo-terminology, you will use it effectively.

→ **How long?** 'How long should my notes be?' is a regular student query; the answer involves lengths of string. The length of notes depends on your purpose. Generally, if notes occupy more than one side of A4, or one to five per cent of the text length, the topic must be of crucial importance. Some tutors will have apoplexy over the last statement; of course there are cases where notes will be longer, but aiming for brevity is a good notion. Look at **Try This 6.1**.

→ **Sources** – keep an accurate record of all research sources and see Chapter 16 for advice on how to cite references.

→ **Quotations** – a direct quotation can add substance and impact in writing, but must be timely, relevant and fully integrated; always ensure quotations are in '…' and are fully referenced in the format: Author date: page number, e.g. Bloggs 2020:432. A quote should be the only time when your notes exactly copy the text. Reproducing maps or diagrams is a form of quotation

Plagiarism and ethical behaviour

The environment is everything that isn't me.

(Albert Einstein)

(BUT when and where did he say this? Is this plagiarism?)

Most students are aware of plagiarism well before university. This chapter covers the key points but please, PLEASE read your university website for guidelines and support. This chapter makes some points about being ethical and honest as a researcher. It may seem obvious but making up or changing data to present a 'better' result is wrong.

> *At school you found something and copied it, and got high marks. How much do you have to change to be OK?*

> *It's where you don't say who wrote something that you include in an essay.*

8.1 What is plagiarism?

"Plagiarism is presenting or submitting someone else's work (words or ideas), intentionally or unintentionally, as your own. This is considered to be a form of cheating and may be subject to disciplinary action, so it is essential to recognise and avoid it."

(Oxford Brookes University, 2011)

"Plagiarism is defined as presenting someone else's work as your own. Work means any intellectual output, and typically includes text, data, images, sound or performance."

(University of Leeds, 2005)

"Plagiarism is the use of another person's words or ideas as if they were one's own. It may occur as a result of lack of understanding and/or inexperience about the correct way to acknowledge and reference sources. It may result from poor academic practice, which may include poor note taking, careless downloading of material or failure to take sufficient care in meeting the

required standards. It may also occur as a deliberate misuse of the work of others with the intent to deceive."

(Flinders University, 2011)

Tutors will conclude that your work is plagiarised when:

✗ it contains sections of text, diagrams or photographs without reference to their original source;

✗ a source is cited but there are no 'quotation marks' around direct (word for word) quotations;

✗ it has been bought from an internet website;

✗ it has been written by another student.

Essentially avoiding plagiarism is about behaving ethically as a researcher. Plagiarism is cheating. Ideas are all around us, university students and researchers are working with these ideas all the time. The important thing is to acknowledge where ideas have come from, and work with them, putting ideas into your own words and framework. This is where the 'thinking' part of research is important.

What is OK? Summarising ✓, paraphrasing ✓, "quotations" ✓ and copying ✗

You will be reading and citing research throughout your degree. The trick is to develop good habits when reading and making notes so that you do not plagiarise. Good practice involves reading a paper or chapter, and then **summarizing** it in your own words. This summary will usually be a shorter version of the original with all the important points included.

Read → Think → Summarize

Paraphrasing involves rewriting something in your own words to convey the same message as the original. Just changing a couple of words in a paragraph is not paraphrasing.

Quotations are fine provided they are placed in "quotation marks". Quotations should be short and verbatim. Use them when the author's words are particularly important. To be clear that you are quoting someone exactly, inset the quotation from the normal margin, or use *italics*.

Working with others is absolutely fine, and an important part of group work. If the assessment is a group report, sharing and writing together is important. If you are asked for an independent report then make sure you work together on the project and notes, but write your own report in your own words. Otherwise you may be accused of **collusion**, stealing your colleagues' work.

Quoting too often is also a problem. The student who submitted an essay where 90 per cent of the words were correctly cited quotations in quotation marks failed the essay because he had not demonstrated his understanding of the material and had simply cut and pasted other people's words. In my experience markers expect less than 5 per cent of an essay to be quotations. Your summaries or paraphrases show you have thought about and understood the material, they demonstrate your own insights and get good marks. Have a go at **Try This 8.1.**

TRY THIS 8.1 – What is OK?

Look at each of these examples in turn and decide what is acceptable academic practice. Then look at the answers at the end of the chapter.

1 Copying three sentences, changing their order, with no acknowledgement and no quotation marks.

2 Taking 200 words from a journal article, and putting it in the essay without any acknowledgement or quotation marks.

3 Paraphrasing the ideas from a paper in your own words, adding information from another source, and acknowledging both in the text, and the references at the end.

4 Copying three sentences from a textbook, changing two words, with no acknowledgement and no quotation marks.

5 Copying three paragraphs from a journal article, putting them in quotation marks, with the authors name and date, and the full citation at the end of the essay.

Ethics and plagiarism

Behaving ethically is essentially about being fair and honest with everyone you meet. This includes being honest and giving appropriate acknowledgement to others in your research and learning activities. It involves behaving in a way that does not infringe the rights of other people, treating everyone with respect and dignity.

There are courses in ethics, or parts of modules which address ethics and ethical behaviour in the GEES degrees. It is especially important in dissertation work to ensure that any research undertaken that involves people or animals respects their rights as individuals, and is also appropriate. Getting ethical approval for a research project is a normal part of the research process.

The Cambridge University (2011) statement on plagiarism considers plagiarism to be a 'breach of academic integrity'. It points out that everyone in the university whether an academic or a student is part of the community which collectively behaves with 'intellectual honesty and transparency'. It talks about the consequences for an individual who plagiarises in terms of the individual's reputation and potential to damage their career. I suggest that you look now at the Cambridge statement and see to what extent this paragraph has summarized or plagiarised the main points.

The Dartmouth College (2008) advice focuses on academic integrity and the nature of research being part of the conversation amongst academics across institutions and generations. 'When you demonstrate that you have done the research required to qualify you to join the conversation, you not only show respect for others' work, you also confer authority upon yourself and highlight the novelty of your particular contribution to the set of ideas under discussion. In these ways, citing sources represents a fundamental step in developing a scholarly voice'. (Dartmouth College, 2008.) How would you summarise this direct quotation yourself?

What should be clear from every website you refer to for advice on plagiarism is that ignorance of the process is not an acceptable excuse. Every student is expected to understand the importance of being honest in all their academic work and that includes referring to other people's work. If in doubt ask your tutor, and include more rather than fewer sources. First year is a good time to make sure a tutor has properly explained the university expectations and procedures to your class.

⇨ TOP TIPS

✓ Every time you make notes ensure you include the full reference so you can put it in your final report or essay.

✓ If you put a quotation in your notes, make sure to put quotation marks "..." round it and highlight it, together with the full reference, so that you will remember to cite it properly in your essay.

✓ Use EndNote to create a database of references (see Chapter 16).

✓ Start working on a report of essay as soon as possible. Summarizing well in your own words is only possible IF you have time to read and then think about things in advance. You need to give your brain time to put ideas into a sensible order and come to some personal conclusions. (People with very little time tend to copy or change a few words, because they don't have time to go through the summarizing process properly.)

✓ Go to your University website and try the plagiarism tutorial, or the activities in **Try This 8.2**. Just do it, make sure you understand, and then you won't need to worry.

TRY THIS 8.2 – Plagiarism activities online

Many universities have exercises to show students exactly what plagiarism involves and how to avoid it. Take half an hour to check out any of the following sites in addition to your own university's plagiarism resources. All sites accessed 5 February 2011.

Indiana University 2010 How to Recognize Plagiarism, https://www.indiana.edu/~istd/plagiarism_test.html

Learn Higher 2010 Preventing Plagiarism, http://learning.londonmet.ac.uk/TLTC/learnhigher/Plagiarism/index.html

McAskill, A 2010 Research Ethics, http://www.socscidiss.bham.ac.uk/s8.html

University of Essex 2010 Authorship and Plagiarism, http://www.essex.ac.uk/plagiarism/test.html

University of Leeds 2010 Plagiarism, http://skills.library.leeds.ac.uk/avoiding_plagiarism.php

University of Toronto 2010 Plagiarism Self Test, http://www.ecf.utoronto.ca/~writing/interactive-plagiarismtest.html

York University 2010 Forms of Academic Dishonesty, http://www.yorku.ca/tutorial/academic_integrity/acaddishforms.html

8.2 Websites – what can I use?

You cannot just assume that information posted to the web can just be used in any way you wish. Start by assuming that you cannot reproduce anything directly. Then look at what might be possible.

Many websites tell you what is copyright and what can be used, but the information is on the font or home page of the website, and/or at the foot of the page.

See for example:

Google	No specific policy as Google has many services
Google Maps and Google Earth	http://www.google.com/permissions/geoguidelines.html
Google Images	http://www.google.com/support/websearch/bin/answer.py?hl=en&answer=9299
Wikipedia	http://en.wikipedia.org/wiki/Wikipedia:FAQ/Copyright#Can_I_reuse_Wikipedia.27s_content_somewhere_else.3F
NASA	http://www.nasa.gov/audience/formedia/features/MP_Photo_Guidelines.html
Greenpeace	http://www.greenpeace.org.uk/help/copyright
RSPB	http://www.rspb.org.uk/help/copyright.aspx

Wikimedia Commons	http://commons.wikimedia.org/wiki/Commons:Reusing_content_outside_Wikimedia
YouTube	http://www.google.com/support/youtube/bin/answer.py?hl=en&answer=83766
Flickr and Creative Commons	http://www.flickr.com/creativecommons/
Yahoo!	http://info.yahoo.com/copyright/us/details.html

8.3 How do lecturers spot plagiarism?

Academic staff read the literature for their subject all the time. In addition they mark hundreds, in some cases thousands, of pieces of student work, every year. They are very aware of writing styles as well as content. The easiest plagiarism to spot is when someone new to a particular topic is writing in an unsure manner. If there are perfectly written, beautifully fluent paragraphs somewhere in a muddled report or essay, they are likely to have been written by somebody else.

> I really enjoyed reading the middle three pages of her report. It was elegant and well referenced. Then I remembered writing it myself, word for word, two years ago. Shame really, she had been a nice student.

Plagiarism is spotted when people use unusual words. It is particularly easy to identify when students copy directly from their lecturers own books and papers. "Always put direct quotations in quotation marks, followed by the author, date and page numbers". (Kneale 2011: 81.)

Remember to be ethical in all your work. This is not just an essay and report problem. Plagiarism will be found in practical reports, maths and stats exercises, and fieldwork reports and mapping exercises.

Many academic departments use software to identify plagiarism. You will be introduced to this process, and usually have the opportunity to submit your writing for checking before submission. If your department uses Turnitin (2011) you will know how to submit coursework. I strongly recommend you go to the workshop on using Turnitin properly.

Diagrams, maps and photographs without references are particularly easy for a tutor to spot, especially on PowerPoint presentations (see Figure 8:1).

A UK Ordnance Survey (OS) map for example is cited as:

Ordnance Survey, 2009. Isle of Man, Landranger Series Sheet 95, 1:50000, Ordnance Survey, Southampton

A photograph should acknowledge the photographer's name with the figure, and the full reference at the end. This may include: Year of production. *Title of image*, and where it is located, for example a library or gallery location:

Davis, J. 2011 Personal photograph, Jura field class, 30 September 2011

For further guidance see Chapter 16 or the Wikimedia Commons (2011) website where there are thousands of images that can be used freely with the referencing details attached to each image.

Diagrams copied or scanned from a paper or chapter need the reference in the usual way with the page number for the diagram included. For example if you copy Figure 2 on page 675 from Prentice *et al.* (2010), you need to include the citation with the figure and page number from the original paper in your essay: From Prentice *et al.* 2010, Fig 2: 675. Your reference list will cite the paper in full:

Prentice CS, Weber JC, Crosby CJ and Ragona R 2010 Prehistoric earthquakes on the Caribbean–South American plate boundary, Central Range fault, Trinidad, *Geology*, 38,8, 675–678, Downloaded from geology.gsapubs.org Accessed 15 February 2011

See Chapter 16 for further details on referencing.

8.4 What happens when I get caught?

You won't be IF you never steal and repeat someone else's work. BUT if you are daft enough to cheat the penalties are severe.

> Passing off somebody else's work off as your own is wrong. Expect to get caught and expelled.

Each university has its own regulations: check your university website and your student handbook. Expect a school investigation, followed by a university process. At each stage you will be given every opportunity to present your case, the Students' Union is usually well-equipped to give advice.

> Arguing that you did not understand what plagiarism is, is not an acceptable excuse.

A first offence from a first year student may just possibly be treated leniently if tutors are persuaded that you were just getting used to the university system. Penalties typically range from repeating the work for the minimum pass mark, repeating the work to pass standard for a mark of zero, or being required to complete and pass the module but get a mark of zero overall. Whatever the penalty, it will affect your final degree marks, which does not look good on your graduation transcript. In second and third year, and for second offences, penalties will range from loss of marks for the module to expulsion from the university.

Whatever the penalty your university imposes it is not worth getting into that position. Look up the message on plagiarism at Johns Hopkins University (2011) which talks about its 'zero tolerance policy'. Similarly from Flinders University (2011):

> "Plagiarism is the use of another's ideas or words as if they were one's own. Plagiarism constitutes academic dishonesty, and is taken very seriously by both the School of Humanities and the University. The penalties for plagiarism may include zero marks for the relevant piece of work, a fail grade for the whole topic, or referral to the Vice-Chancellor."

 TOP TIPS

Avoid all the hassle involved in being suspected of plagiarism by:

✓ Starting assignments early;
✓ Practising writing summaries;
✓ Using quotations properly with "..." and adding complete references;
✓ Checking carefully that you have not copied;
✓ Running your work through Turnitin before submission if this is available to you.

8.5 Ethical behaviour and data

Collecting data, running experiments and field-work all take time. High quality GEES research is produced by people thinking hard about their tests or experiments, and who are rigorous about the quality of the data they collect. Making up data is cheating. Universities take a dim view of students who cheat. Students who

> **Cheating** is trying to gain an unfair advantage by attempting to deceive tutors by making up data or presenting other people's data as your own in coursework or exams.

are caught cheating have usually resorted to it because they are short of time to do their work properly. This is why dissertations are usually six- or nine-month activities. This allows time to design and collect data to a high standard, to analyse the data, to think about the results and what they mean.

Why might these graphs ring alarm bells for a marker?

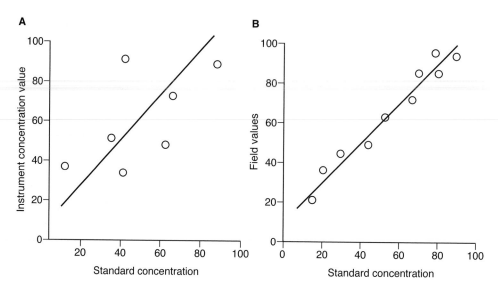

Figure 8:1 A Laboratory calibration of instrument against standards; B Range of results recorded in the experiment or the field

In this figure the left-hand graph (A) shows the laboratory calibration for an instrument against a standard, the right-hand graph (B) shows the field results. Where an instrument has that much variation when tested with laboratory standards, it is bound to be recording with greater variability in the field. People who cheat tend to present data that is too perfect, too good to be true.

It can be really hard to get people to fill in questionnaires, especially if you're standing in a town centre and it's raining. The temptation to get 15 answers and make up the next 15 is high. Please don't do it. Tutors will ask to see the raw data, look at the handwriting, and will notice if amazingly you did five questionnaires in five hours on three days, and then completed 40 questions on day four. Tutors realise how difficult it is to get good quality data and answers.

If you have limited numbers of responses ask your tutor whether you should request an extension to do further research, or explain why your data is limited in your report. Be cautious about the interpretations you make from it. Cheating by falsifying data just isn't worth considering.

It is normal in the data set to find anomalous results. These come from human error in collecting the data, errors in instrumentation, errors in entering the data for analysis, and … most field values are highly variable, with hourly, daily and seasonal variations, as well as point to point variability. GEES data is always going to be messy. The fun is in interpreting the muddle. Generating completely clear conclusions is likely to ring warning bells for markers. If there are some anomalous data points comment on them, please don't delete them. It is proper to suggest you might have made an error in data collection or data entry. It may be there is some variation of which you are unaware, but which your tutor expects. For example the student who replaced his recorded pH data at station x because it was out of line with all the other values, had missed the change in rock type at that point. The tutor expected to see very different pH levels at that station and identified the cheating.

Always keep all your data sheets, questionnaires, notebooks, etc. until after the research is finished, reported and marked so that you can check back to your original data.

In some departments there are student research projects where data are shared across years; your dataset includes your results and results from students in previous years. Then your results become part of this larger dataset for next year's students. Keeping accurate records is obviously very important in this case, but equally important for all projects. Being honest in reporting results shows sound judgement as a scientist or social scientist. You want this reputation.

8.6 References and resources

Blum SD 2009 *My Word! Plagiarism and College Culture*, Cornell University Press, Ithaca

Burkill S and Abbey C 2004 Avoiding Plagiarism, *Journal of Geography in Higher Education*, 28, 3, 439–446

Cambridge University 2011 Plagiarism – Information for students, http://www.admin. cam.ac.uk/univ/plagiarism/students/index.html   Accessed 15 February 2011

Dartmouth College 2008 Sources and Citation at Dartmouth College, http://www. dartmouth.edu/~writing/sources/sources-citation.html   Accessed 15 February 2011

Flinders University 2011 Policy on Academic Integrity, http://ehlt.flinders.edu.au/ humanities/exchange/style/integrity.html   Accessed 15 February 2011

Johns Hopkins University 2011 Plagiarism, http://advanced.jhu.edu/students/ plagiarism/   Accessed 15 February 2011

University of Leeds 2010 Plagiarism – University of Leeds Plagiarism Guide, Office of Academic Appeals & Regulation, http://www.ldu.leeds.ac.uk/ plagiarism/wphp   Accessed 15 February 2011

Oxford Brookes University 2011 Plagiarism – what is it? http://www.brookes. ac.uk/library/skill/plagiarism.html   Accessed 15 February 2011

Turnitin UK 2011 Turnitin2, http://www.submit.ac.uk/static_jisc/ac_uk_index. html   Accessed 15 February 2011

Turnitin 2011 Prevent plagiarism. Engage students http://turnitin.com/static/ index.html   Accessed 15 February 2011

Wikimedia Commons 2011 Welcome to Wikimedia Commons, http://commons. wikimedia.org/wiki/Main_Page   Accessed 15 February 2011

Answers to **Try This 8.1**: Points 1,2 and 4 are absolutely unacceptable plagiarism. Point 3 is OK. Point 5 is lazy academic work. It may not be deemed to be plagiarism, but expect tutors to give you a low mark for not thinking through and developing your own take on the information. If you find you have embedded a long quotation, the advice is to paraphrase and re-present it in your own words.

Geographs 2

Find the three GEES terms in these jumbled letters. Answers on p.297

9

Creativity – Innovation

Reading a book is a really bad way to engage with creativity and innovation, therefore this chapter is very, very short. Follow the web links. What does creativity mean for you? What does creativity mean in the earth sciences, in geography and in environmental sciences?

> **Creativity** – generating a new product or process.
> **Innovation** – developing a product or process to a new level.

These questions are not easy to answer. If you want your university experience to involve getting a degree through learning stuff and then getting a graduate job, creativity may not be important for you. There is a view that university is about education in its broader sense of helping students to develop their full potential. Being creative is part of that process, so that higher education makes a positive difference to your life.

Sternberg and Lubart (1995) argued that combining three groups of abilities or attributes led to success:

- *analytical attributes* – analysis, evaluation, compare and contrast, devising a project plan …
- *practical attributes* – applying ideas, active research, fieldwork, implementing a project plan …
- *creative attributes* – imagination, synthesis, making new connections, discovering, adapting imagining, inventing …

There is some literature which suggests that creative people who have the opportunity to use their creative attributes are more personally satisfied. But that's another book. It is often easier to identify the analytical and practical elements to be covered in a module from the handbook or course outline, but academics are always looking for the creative

> ***Six Thinking Hats***
> (de Bono, 2010)
>
> White Hat asks for information;
> Red Hat asks about feelings, hunches and intuition;
> Black Hat asks difficult questions and suggests why something won't work;
> Yellow Hat is optimistic, cheerful: it will be okay;
> Green Hat looks for creative possibilities and new ideas;
> Blue hat manages everyone's thinking.

elements. The aim of most academics is to make the new connections, adopt, adapt and imagine new ways of doing their particular part of the subject. Projects, essays and dissertations are usually designed so that students can develop and demonstrate these creative attributes.

One good place to start thinking about what creativity might mean for you is with Edward de Bono. He has written various books and you can hear his ideas about creativity on YouTube. Google 'YouTube and de Bono with creativity/six thinking hats/problem solving' as a starting point.

Essentially creativity for de Bono is about using your imagination to develop new ideas and new ways of doing things, thinking laterally. It is not about change, and making changes and developments for their own sake. Creative thinking leads to change where there are good reasons for change.

> Putting in my ideas seemed really weird until my tutor explained how boring it is to mark 280 essays all saying what he said in the lecture.

In his Six Thinking Hats activities, people take different roles. The Green Hat person offers new ideas, encourages brainstorming, and thinking beyond the usual solutions. (de Bono 2010, Mind Tools 2011.) Aim to find more times to practice wearing the green hat in all parts of your life.

Creativity and innovation are terms used synonymously but which have different meanings. Creativity is about imagining and developing novel solutions. Innovation is about taking an idea or product and developing it further. A geologist imagining a completely new way to extract oil would be thinking creatively. A geologist improving an extraction technique by developing the technology is being innovative.

Being creative or innovative requires action. Joel Barker, an American businessman is regularly quoted:

'Vision without action is merely a dream.
Action without vision just passes the time.
Vision with action can change the world.'

Universities and university research are about educating students and solving research problems through research. University academics are interested in understanding how the world works and inventing solutions to problems and issues for society. So being creative and innovative are core skills for all university lecturers.

SCAMPER (a checklist for changing perspectives)

Substitute – seek alternatives to products, processes or roles.
Combine – mix and integrate different approaches or activities.
Adapt and adjust – change the expected process or procedure.
Modify the current process or product.
Put to new uses – finding new applications.
Eliminate – simplify by removing unnecessary functions.
Rearrange or reverse – reorder process activity or items.

Developing creativity skills means exercising your brain. Use **Try This 9.1** and apply the ideas in your university work. Success is making connections between concepts and ideas and presenting them with academic evidence (academic evidence means a good, reasoned and logical argument with at least *two references* and possibly more). There are various websites that prompt creative approaches to research thinking, and creative thinking in your everyday life. See **Try This 9.1.** for some starting points. SCAMPER and Six Thinking Hats are just two ways to approach creative thinking.

TRY THIS **9.1 – Creativity and me**

These websites are good starting points. Take half an hour to explore them, then have a think about your own approaches to problem solving, thinking around issues, and your own characteristics – confidence, curiosity, ability to be persistent, for example.

Creax (2011) This site is aimed at people in work. Adapt it using 'Others' as your response to the last question on page 1, and then reply as if the main question starts 'On a typical day at university …' Or do the test thinking about your part-time or vacation job. www.creax.com/csa/

Mind Tools (2011) is an excellent site. Start with the Practical Creativity section – Developing Creative Solutions to the Problems You Face, SCAMPER and Just Do it are great starting points. www.mindtools.com/pages/main/newMN_CT.htm

Creativity in the Biosciences (2011) – this is not GEES-related but the site is excellent and the activities work beautifully for all GEES students, social scientists or scientists. Register for a username and try out the activities. If you don't know where to start try the Idea Dump, SCAMPER or Googlestorming. www.fbs.leeds.ac.uk/creativity/login.php

---

Most people agree that being creative is a group activity – brainstorming, idea dumping, mind mapping, exchanging thoughts and concepts are more fun in a group. It is important to understand how people from different backgrounds, cultures and age groups respond to new thoughts, concepts and products. Having creative ideas and finding new ways to do things is important, and so is evaluating them sensibly. For GEES people pilot projects are crucial for evaluating new ways of researching, testing, experimenting and thinking; to be creative or innovative geographers, earth and environmental scientists.

JUST DO IT.

9.1 References and further reading

Adams DJ and Sparrow JC (Eds.) 2008 *Enterprise for Life Scientists*, Scion, Bloxham

Creax 2011 Creativity Self-Assessment Page, http://www.creax.com/csa/   Accessed 15 February 2011

de Bono E 2010 *Six Thinking Hats*, Penguin Books, London

Litemind 2010 Creative Problem Solving with SCAMPER, http://litemind.com/scamper/   Accessed 15 February 2011.

Mind Tools 2011 Creativity Tools, http://www.mindtools.com/pages/main/newMN_CT.htm   Accessed 15 February 2011.

Sternberg RJ and Lubart TI 1995 *Defying the crowd: Cultivating creativity in a culture of conformity,* Free Press, New York.

Can you find the original J. Barker reference? See p.88. Try Google! Like many quotes, this one is cited on many sites. Such quotes are better excluded from university work because you do not know the source.

Add up 1

Work out the totals for the top of the pyramid. Answers on p.297.

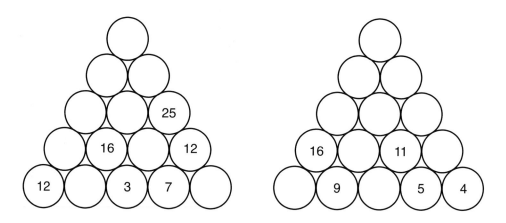

10

Thinking

Most students are too busy to stop and think. Are you?

I don't think I understood what I said either.

Refusing to go out because 'I have to stay in and think about magma generation and crustal anomalies in the lithosphere', or 'the implications of positive discrimination as a factor in the self-perpetuating cycle of poverty in three developing cities', is not cool. The old excuses 'washing my hair' or 'I need a quiet TV night in' are probably better. But creating thinking time is really important. For most people, thinking is started by conversations and discussions. Being asked: 'How do geothermal technologies actually work?' or 'What is your understanding of the Chicago geographers' advocacy of urban ecology as affecting our current understanding of interactions in the city?' or 'How do soil pore size and buffer strips affect crop growth?' can stimulate thoughts you didn't know you had.

Did you put 'thinking time' into your plan for today? OK, Do it now.

Thinking happens all the time. GEES students are expected to apply their already well-developed thinking skills to a series of academic tasks and activities, to make reasoned judgements and arrive at conclusions about issues. All GEES degrees can be studied at a rather superficial level; people learn and re-present information. This is called surface learning and usually leads to marks from 45 to 60 per cent. The aim of a university education is to really understand through reading, discussing, practical field and laboratory work and thinking about the topics; it is called deep learning.

Success involves asking questions, thinking about what you do and don't know, relating ideas and feedback from one module to other parts of your degree and to other subjects. It's all about developing your ability to interrelate evidence and draw valid conclusions. This links to the ideas of deep reading (see p.60).

Environmental Science and Geography degrees obviously involve knowledge from different disciplines, economics, sociology, law, biology, chemistry, mathematics and statistics, and students regularly choose modules from other disciplines as part of the course. Earth Science builds on chemistry, physics, maths and engineering. As a GEES student you have the fun of extracting ideas, researching approaches and information and getting out in the field to develop and support your own thinking about any research idea with evidence. (*Evidence* means *add references*.) You will think more like a geologist, (Manduca and Mogk 2006) geographer or environmentalist.

You will get better at thinking, more intellectually sophisticated, during your degree course just by doing it. This chapter can help you understand what's going on, and help you to reflect on how and when you think. It is a very minimal excursion into thinking-related activities. It ignores most of philosophy and

I'd never thought about how I think, nobody ever told me it was important before. Which is really strange when you think about it. These questions seemed completely mad at first, but began to really help, reading was easier. I was taking better notes

the cognitive sciences and just concentrates on three elements: unpacking what it means to be critical, reasoning, and questions to encourage and focus your thinking. To develop sparkling thinking, start by considering the ideas here. Like bicycle stunt-riding, thinking gets better with practice, not overnight. Thinking is tough, chocolate can help.

10.1 Why do you think?

Thinking is used to acquire understanding and answers, to make sense of our complex world. Adjectives used to describe *effective* thinking include: reasoned, clear, logical, precise, relevant, broad, rational, sound, sensible and creative. You cannot think in the abstract; there must be a purpose to your thinking, even if it is not immediately obvious. The process of good quality thinking will probably involve:

- deciding on your objective (problem solving, understanding a concept, developing a hypothesis)
- defining and understanding the background terminology and assumptions
- acquiring data and information of a suitable standard to build up a reasoned argument
- reasoning or inferring from the available information to draw logical conclusions
- considering the consequences of the results, applying them back to the problem or to a new situation.

You can get better at thinking. Research thinking is helped by tracking the process in your notebook, or lab or field notebook so you can see ideas develop and can follow them through at different times. Most GEES issues are multi-dimensional (carbon capture, heawater stream ecology, geochronology, core-periphery dependency theory), which makes taking notes while you are thinking really important. Try plotting your thoughts on spider diagrams, and record connections and links as they occur to you.

There are apps and online packages which will make your diagrams look smart but handwritten in your notebook, with your research notes, is just fine. It's the thinking and links you want rather than the whizzy technology.

> Ideas float away all too easily.

10.2 Critical thinking

Critical thinking involves working through a problem. This involves starting by thinking about the nature of the problem, thinking through the issues and striving for a reasoned, logical outcome. During the process you need to be aware of other factors that impinge, where bias may be entering an argument, the evidence for and against the issues involved, and to search for links to other parts of your discipline and other disciplines. Earth scientists will check out the latest chemistry thinking while social geographers are reading the sociology literature. Essentially, you are critically evaluating the material throughout the process of researching and thinking about it.

Being critical entails making judgements on information. It is important to remember that being critical does not necessarily imply being negative and derogatory. It also means being positive and supportive.

> **Essay marks**
> Ask a tutor how s/he assigns essay marks. Does the percentage balance suggested here change from Year 1 to Years 2 and 3? What is your critical content worth?
>
> Maybe:
> *1 Essay writing skills 10%* – good introduction, conclusion, paragraphs for each argument, grammar and spelling.
> *2 Information literacy 30%* – use of resources from the reading list 10%, and not on the reading list 20%.
> *3 Critical thinking skills 60%* – logically ordered ideas, good balance of evidenced arguments and case examples, relevant material from a broad range of sources, thoughtful conclusions.

It involves feeding back in a thoughtful way. A balanced critique looks at the positive and negative aspects. Some students feel they cannot make such judgements because they are unqualified to do so. Recognize that neither you, nor your teachers, will ever know everything: you are making a judgement based on what you know now. In a year's time, with more information and experience, your views and values may alter, but that will be a subsequent judgement made with different information.

Discussion is a major thinking aid, so talk about migration, biodiversity on coral reefs and the stratigraphy of Dorset to colleagues, friends and family. It can be provocative and stimulating! You will have new thoughts.

Where does intellectual curiosity fit into this picture? University research is about being curious about concepts and ideas. You can be curious in a general way, essentially pursuing ideas at random as they grab your imagination. More disciplined thinking aims to give a framework for pursuing ideas in a logical manner and to back up ideas and statements with solid evidence in every case.

Avoid uncritical, surface learning, which comprises listening and note-taking from lectures and documents, committing this information to memory and regurgitating it in essays and examinations. You will acquire knowledge this way, but the 'understanding' step is missing. Assignments which are mostly facts (*knowledge*) tend to get half marks. University assessments for higher marks show evidence of deeper learning, evaluation and applications to different contexts.

10.3 Thinking styles

Ways in which people think and learn were described by Bloom (1958) in his 'Taxonomy of Learning'. It has six levels starting with knowledge and comprehension where facts are recalled and repeated. He describes higher order, more complex and abstract levels of learning as progressing from 'application', 'analysis', and 'synthesis' to 'evaluation'. Figure 10:1 summarizes this concept with a series of associated verbs. University learning should be at the higher levels. You are expected to progress from knowledge-dominated activities to those with increased emphasis on analysis, synthesis, evaluation and creativity. Use **Try This 10.1** to make notes about when and how you use your brain for each of these different styles of learning.

Knowledge	Write; state; recall; recognize; select; reproduce; measure.
Comprehension	Identify; illustrate; represent; formulate; explain; contrast.
Application	Predict; select; assess; find; show; use; construct; compute.
Analysis	Select; compare; separate; differentiate; contrast; break down.
Synthesis	Summarize; argue; relate; precise; organize; generalize; conclude.
Evaluation	Judge; evaluate; support; attack; avoid; select; recognize; criticize.

Figure 10:1 Summary of Bloom's (1958) skill taxonomy

⚙ TRY THIS 10.1 – Where have you, and how are you, using the different Bloom skills?

Bloom's Taxonomy of Thinking Skills	Examples of this type of learning	Where and how in my degree, at school, at work?
Knowledge	Listing information, repeating from notes, memorizing a diagram, recognizing ... minerals, rocks, plants, clouds ...	
Comprehension Understanding	Reviewing information, selecting and re-presenting an issue ...	
Application	Applying an idea, concept or process from one area of the GEES disciplines in another area or environment.	
Analysis	Looking at an issue systematically, working out the different parts involved and examining the issues in turn.	
Synthesis	Designing a process with multiple steps, developing a procedure to manage ... Assembling a team with the diverse skills to ...	
Evaluation	Judging which arguments or approaches are most valuable and applying them to a problem. Selecting and evaluating options for solving a ...	

University assessments, typically through essays and examination questions, test the way that students think about and manipulate information. You might find it useful to ask yourself questions to help your thinking. Questions which start with 'What do I know about ...?' will help to marshal information at the lower-level, knowledge, end of Bloom's scale. This can be a good starting point BUT then move to the deeper learning end of the scale with questions that start with:

- What criteria can be used to assess ...?
- Evaluate three potential solutions you would suggest to ...
- What must be considered in order to decide about ..?
- What are the most important decisions to be taken in order to ...?
- How can different opinions about ... be evaluated and reconciled?
- How and why would you decide to ...?
- Evaluate the different dimensions approaches to ...?

I can lead my students to the reading list but I cannot make them *think*.

The trick is to set yourself an imaginary question to answer as you are researching an issue. It can change the way in which you make notes.

10.4 Reasoning

Strong essay and examination answers look at the arguments, draw inferences and come to conclusions. Judgements need to be reasoned, balanced and supported. (*Supported* means *add references*.)

First, think about the difference between reasoned and subjective reactions, and reflect on how you go about thinking. Subjective reaction is the process of asserting facts, of making unsupported statements, whereas reasoning involves working out, or reasoning out, on the basis of evidence, a logical argument to support or disprove your case with at least *two references* (see Table 10:1).

Create examples of reasoned rather than subjective statements with **Try This 10.2**. In your academic thinking and communications, avoid making emotional responses or appeals, assertions without evidence, subjective statements, analogies that are not parallel cases, and inferences based on little information, unless you qualify the argument with caveats.

Subjective statement	Reasoned statement
Ethnic minorities are a problem	Smith (2020) shows that within inner city areas different ethnic groups raise different issues for social service provision. The cultural heritage and lifestyle patterns in contrasting groups means that the response to different sections within the community need to be appropriately tailored.
The Mississippi is a very dirty river	Generally the Mississippi carries a very high sediment load during flood flows because high discharges erode bed and bank material and overland flow entrains sediments from the extensive catchment (McLean and Levy 2020). The sediment load will be radically reduced immediately downstream of any of the Mississippi reservoirs due to within reservoir sedimentation from the slower moving water.

Table 10:1 Examples of subjective and reasoned statements

TRY THIS 10.2 – Reasoned statements

Either write a fuller, reasoned version of the five subjective statements below,
OR pick a few sentences from a recent essay and rewrite them with more evidence, examples and references. For example answers see p 297.

1 The questionnaire results are right.

2 Pelagic sediments are found in the deep oceans.

3 Man is inflicting potentially catastrophic damage on the atmosphere and causing worldwide climate change.

4 The United States has become an urban country.

5 Pedestrianization civilizes cities.

10.5 Questions worth asking

Being a critical thinker involves asking questions at all stages of every research activity (lecture, tutorial, workshop, laboratory, discussion, party). These questions could run in your head all the time. Pick a couple and practise with them each day:

- What are the main ideas here?
- Are the questions being asked the right ones or are there more meaningful or more valid questions?
- What are the supporting ideas?
- What opposing evidence is available?
- Is the evidence strong enough to reach a conclusion?
- How do these ideas fit with those read elsewhere?
- What is assumed?
- Are the assumptions justified?
- What are the strengths and weaknesses of the arguments?
- Is a particular point of view or social, scientific or cultural perspective biasing the interpretation?
- Are the definitions right?
- Is there evidence of high technical standards in the analysis?
- Is the data of an appropriate quality?
- Do the results really support the conclusions?
- Are causes and effects clearly distinguished?
- Is this a personal opinion or an example of intuition?
- Have I really understood the evidence?
- Am I making woolly, over-generalized statements? 'There is a serious problem with housing in America'. OK, but 'In the inner cities of the north-eastern United States there is under-provision of basic housing' (Penn and Ink 2020), or 'The extensive development of retirement housing in Florida has taken a disproportionally high percentage of the state budget for 2015 (Elder and Disney 2020)', are better.
- Are the points made/results accurate? 'Plate margin movements are more active now than at any other time in the last 5000 years' – sounds clear, but is it right? Where is the evidence? Where are the references?
- Are the results/points precise? 'The Ganges floods annually'. This statement is clear and accurate, but we do not know how precise it is. How often does the river flood? Where does it flood?
- Is this information relevant? Keep thinking back to the original aims and argument. You can make statements that are clear, accurate and precise, but if they are irrelevant they do not help. Off-the-point arguments or examples distract and confuse the reader, and may lose you marks.

- Is the argument superficial? Have all the complexities of an issue been addressed? 'Should slums be swept away?' A clear 'yes because ...' will answer the question, but this is a very complex question, requiring consideration of the economic, social, cultural, historical, political and planning perspectives. Ask what is meant by a slum – it has different connotations in Colchester and Calcutta.
- Is there a broad range of evidence? Does the answer take into account the range of possible perspectives? The question 'Discuss the arguments for and against climate change' is so broad. You cannot cover all points in equal depth, but aim to make the reader aware that you appreciate there are more points of view or approaches.
- Are the arguments presented in a logical sequence? Check that thoughts and ideas are ordered into a sequence that tells the story in a logical and supported way.
- How does this idea or hypothesis fit with the wider field of enquiry? You might be looking at a paper on the incidence of asthma in Indonesian communities, but where does it fit with medical geography, population geography or regional studies?
- Which examples will reinforce the idea?
- Can this idea be expressed in another way?
- What has been left out? Looking for 'gaps' is an important skill.
- Is this a definitive/true conclusion OR a probable/on-the-balance-of-evidence conclusion?
- What are the exceptions?

TRY THIS 10.3 – Thinking about a paper

Select one paper from a reading list, any paper, any list! Make notes on the following:

Knowledge/Content:	What are the main points?
Evidence:	What is the support material? Is it valid? Are there references?
Counter case:	What are the counter-arguments? Has the author considered the alternatives fully and referenced them? What Bloom (1958) level is being used?
Summary:	Summarise relevant material from other sources that the author might have included but omitted.

How well did the author meet his or her stated objectives? What research could be done next to fill in gaps or follow up on this study?

⚙️ **TRY THIS 10.4 – Comparing papers**

Take three papers that are on the same or related topics, from any module reading list. Write a 1000-word review that compares and contrasts the contributions of the three authors. (Use the guidelines from the previous exercise.) Write 250 words on where these three papers fit with material from the module.

This will take time, but it really will improve your comprehension of a topic. Treat it as a revision learning exercise. Pick three papers you are going to read anyway.

Take time to think and reflect before jumping into a task with both feet. Having completed a task or activity, take a few minutes to reflect on the results or outcomes. **Try This 10.3** and **10.4** are tutorial exercises that develop critical thinking skills. Both provide frameworks for thinking, evaluating and synthesizing.

In practice, few lecturers would argue that a logical perspective is the only way to deal with questions. Express your aesthetic opinions within an essay, if they are relevant and appropriate. An essay on 'the nature of flow processes in rivers' requires equations for fluid flow, Reynolds numbers and case examples. A poetic answer describing the delight you feel in viewing a flowing stream at dawn will not do, and you will be in danger of being deemed 'a tributary short of a river'.

Can you improve the quality of your thinking alone?

Yes, but it takes practice. You will become more disciplined in your thinking by discussing issues regularly. This is because the act of talking around an idea sparks off other ideas in your own mind. When someone else voices their point of view, you get an insight into other aspects of the problem. Thinking of arguments that run against your own position is difficult. A discussion group might do the following:

✓ start by summarizing the problem
✓ sort out objectives to follow through
✓ share data and evidence, the knowledge element
✓ share views on the data, 'I think it means ... because ...'
✓ work out and discuss the assumptions the data and evidence are making
✓ discuss possible implications; evaluate their strengths and weaknesses
✓ summarize the outcomes.

> I had a great Year 1 tutor, lovely, he talked all the time. I hated my Year 2 tutor, because he wouldn't tell us anything. But that turned out to be really useful. We had to work it out ourselves.

A good reasoner is like a good footballer: it takes practice.

Thinking takes time

> I knew my tutorial group were winding me up by saying 'We don't know – what do you think?' But that's not the point. I do have an answer, but the group needs to learn how to get to an answer. I asked them to take 15 minutes to discuss, make notes and put their ideas on a poster. Then they had lots to say. It just takes time for thinking to happen.

Your brain needs time to organize information and find conclusions. A good 'discussion with thinking' conversation will have long gaps while people organize their ideas. Good tutorials have quiet tutors who ask a question, and then wait for people to offer answers, and to develop the answer in more depth. A tutorial is *your* thinking time, not 'teaching' time! Give people time to answer when you are discussing uni 'work stuff'.

Where to think?

Thoughts and ideas arrive unexpectedly and drift off just as fast unless you make a note. Take a minute to recall where you do your thinking. There are almost as many varied answers as people, but a non-random sample of individuals in a lecture (N=89) shows favoured locations include in bed at 5 a.m., on the Wii, walking to work, running, swimming, at the gym, cleaning the kitchen or cutting grass. There is evidence of thinking being most productive while you are doing something that allows the mind to wander in all sorts of directions, without distracting phones and conversations. The majority of students who offered 'walking' and 'the gym' as their best thinking opportunities are evidence of this. Writing down ideas, or dictating them to your phone immediately, is vitally important. Take 10 minutes over a drink after exercise, or a couple of minutes at a bus stop, to jot down new thoughts and plans. This makes aerobic exercise an effective thinking, multi-tasking activity.

Avoiding plagiarism

Good thinking habits can minimize your chance of inadvertently plagiarizing the work of others. Get into the habit of engaging and applying concepts and ideas, not just describing or reporting them. That means thinking around the ideas to find your own contexts and alternative examples. Make sure you include your own thoughts, opinions and reflections in your writing. Be prepared to draft and redraft so that the thoughts are in your own language, and acknowledge your sources (*add references*). Leave time to link ideas coherently. Finally, put the full reference for your citations at the end of each assignment.

Thinking and understanding involves a commentary in your head. Writing a summary in your own words is a good way to check you understand complex ideas. Ask: 'Do I understand this?' at the end of a page, chapter, paper, tutorial, lecture ... and not just at the end.

10.6 References

Bloom BS (Ed.) 1956 *Taxonomy of Educational Objectives: 1 Cognitive Domain*, Longman, London

Buzan T 2003 *Use Your Head: Innovative Learning and Thinking Techniques to Fulfil Your Potential*, BBC Books, London

Buzan T 2009 *The Mind Map Book: Unlock Your Creativity, Boost Your Memory, Change Your Life*, BBC Books, London

Manduca CA and Mogk DW (Eds.) 2006 Earth and Mind: how geologists think and learn about the earth, *The Geological Society of America*, Special Paper 413, Boulder, Colorado

Richard PW and Elder L 2002 *Critical thinking: Tools for taking charge of your professional and personal life*, Financial Times Prentice Hall, New Jersey

Van den Brink-Budgen R 2000 *Critical Thinking for Students: Learn the Skills of Critical Assessment and Effective Argument*, 3rd Edn., How To Books, Oxford

Geo-codeword 2

Replace the numbers with letters starting with the three indicated below. Complete the grid to find GEES related terms. Answers on p 298.

1	2	3	4	5	6	7 C	8	9	10	11 T	12	13
14	15	16	17	18	19	20 S	21	22	23	24	25	26

	21	16	21	11	4	13		11	21	19	9	26
1		21		4		9		18		15		9
26	21	8	4	6		15	25	4	6	26	21	2
3		15		6		12		6		26		9
11	20	3	5	21	19	9		19	15	3	5	13
4				7		5				20		
5	9	7	14	4	26		9	7	4	7	21	2
		9				24		15				21
10	21	11	7	18		4	22	3	21	11	15	6
4		9		15		11		5		6		20
26	9	17	21	6	13	20		11	6	9	16	4
20		4		20		21		6		21		7
18	4	5	1	4		19	21	8	23	26	8	

Constructing an argument

Two geographers walking through the city observed two people shouting across the street at each other. The first said, 'Of course, they will never come to an agreement'. 'And why?' enquired his mate. 'Because they are arguing from different premises'.

At the centre of every quality GEES essay, report or presentation is a well-structured argument. Good arguments establish a set of facts, the connections that link them together, and a series of examples so that the reasoning and conclusions appear probable. You are not aiming to provide mathematical proof, or sufficient evidence to make a legal case. You are attempting to establish enough links, and to supply the case evidence through geography, earth or environmental examples, so that your argument cannot be rejected as improbable (see Bonnett 2008).

> **Argument:** a fully supported and *referenced* explanation of an issue.

Start by thinking about your motive for putting pen to paper, not just 'I want a 2.1'. What do you want to say? Who is the audience? Most GEES issues can be viewed from a number of aspects, and exemplified with a wide range of material, possibly with evidence from other disciplines such as economics, physics, psychology, history. This means unpacking the elements of an argument and structuring it logically. Your aim might be to:
- develop a point of view, as in a broadsheet newspaper style editorial or a debate
- persuade the reader, possibly your examiner, that you know about 'post colonial changes in agriculture in the developing world' or 'the role of spatial modelling to forecast the spread of pollutants'.

You may be more specific, perhaps wanting to define:
- how one concept differs from the previous concepts of ...
- the implications of using an idea or concept rather than another one
- whether a decision is possible or 'Are we sitting on the fence?'
- what would need to be adapted or amended, what would work and what would not, if an idea or concept was transferred to a new environment.

Some arguments are linear, others rely on an accumulation of diverse threads of evidence which, collectively, support a particular position. Recognize that

Feedback in discussions comes from everyone in the group, all through the discussion. Use their thoughts about your ideas to further develop your ideas. Be open to their ideas.

different researchers will present equally valid, but conflicting, views and opinions. In considering either transport issues or mineral extraction in Devon, you could seek information from ecologists, civil engineers, agricultural scientists, transport experts, political scientists, local councillors, employment experts, environmental campaigning groups, heritage protection

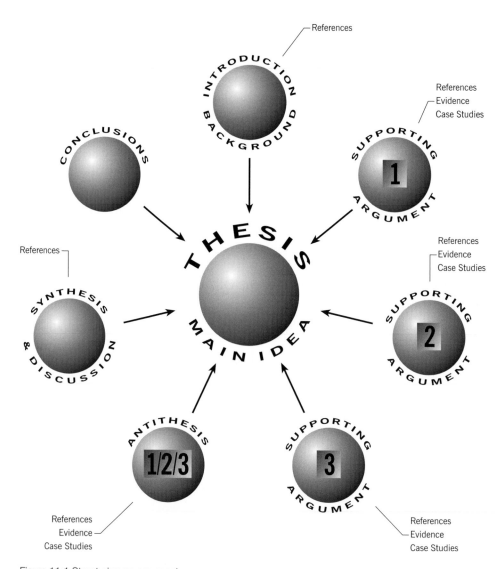

Figure 11:1 Structuring an argument

organizations, local people, road users and many more. In developing a statement about rural transport, you would need to present the different views and consider:
- What are the limitations of each view?
- What elements are entrenched?
- How might a consensus be formed?

If the article is for a newspaper or radio interview, its style should be very different from an examination essay (Hay and Bass 2002).

A GEES research project can leave you overwhelmed with evidence. 'Thinking' is involved in designing an approach to maximize your understanding of influences and interactions between different facets and elements of the data and presenting these in a coherent argument. Points of view should be part of the chain in an argument, or arise from the argument. The weight of the evidence or ideas makes the point of view acceptable.

11.1 Structuring arguments

Any argument needs to be structured (see Figure 11:1), with reasoned evidence supporting the statements. A stronger, more balanced argument is made when examples against the general tenet are quoted. Russell (2001) describes three classic structures that can be adopted in arguing a case (Figure 11:2). It is this author's contention that at university level, the third model should be used every time. In every discussion, oral presentation, essay, lab and field report, and dissertation you should be able to point to each of the six sections and to the links between them.

1	Introduction	Thesis	Conclusion	References		
2	Introduction	Thesis	Antithesis	Conclusion	References	
3	Introduction	Thesis	Antithesis	Synthesis	Conclusion	References

Thesis: the argument being considered
Antithesis: arguments against the thesis
Synthesis: balancing the various points made in favour of and contradicting the argument

Figure 11:2 Models for presenting arguments (after Russell 2001)

11.2 Unpacking arguments

It is all too easy when speaking and writing to make very general statements. One might say, 'Poverty in Africa is a consequence of the historical political systems'. This is true, but hides much information. A fuller statement like, 'Inequality in Africa has complicated historical origins. The roots of poverty in contemporary Africa can be traced back to earlier historical periods, particularly colonial times, and can be seen as a consequence of the unequal relations that exist between "North" and "South". It has also been argued that inequality in Africa is a consequence of the emergence of a capitalist world economy'.

Your level of explanation depends on context. Be careful not to explain the totally obvious. A couple of examples:

- 'Water is a liquid with chemical composition H_2O and flows downhill'. At university level we can assume people know what water is.
- 'The Richter scale measures earthquakes'. This is true but not detailed enough. 'The logarithmic Richter Scale measures the energy released when the ground moves due to seismic activity. It is scaled from 0–10 although the largest recorded event to date is under 9'. This has additional detail and would be excellent as a starting sentence in a Level 1 essay. If you are doing a final year module 'EARS3100 Earthquakes, Seismology and Management', this could be information that is already assumed to be held in common. If in doubt keep the technical content high.

Another statement meriting some unpacking is, 'Men and women play different social roles in developing countries'. A more detailed statement might say, 'Gender is an important factor in the practice and experience of development. In many societies, men and women have clearly defined and often distinct roles. The different tasks assigned to men and women in each society comprise an important part of their gender identities'.

From the soils area, 'Microbiological processes control plant nutrient uptake'. This is a straightforward sentence and might be an essay title, just add 'Discuss'. A version gaining more marks by adding detail might read:

'Soil micro-organisms, which include bacteria, mycorrhizal fungi, actinomycetes, protozoa and algae, control a number of soil processes, recycling and releasing nutrients for plants and other soil organisms. The presence or absence of oxygen determines the species present and their activity rates. The mycorrhizal fungi, for example, are symbiotic dwellers on host plant tissues, taking C from the plant and making P and other nutrients available for plant uptake'.

When reviewing your own writing, first drafts of essays and reports, identify 'general' sentences which would benefit from a fuller explanation.

Q. Did you spot the major flaw in these examples?
A. There are no embedded references. To further strengthen each example add references. No references equals marks below 50%.

11.3 Rational and non-rational arguments

Arguments are categorized as being rational and non-rational. The non-rational are to be avoided wherever possible. Here are five examples:

- Avoid bold statements like 'Rio de Janeiro has the worst/best/smallest/ greatest ... slums/geology/water supply/housing policies/landslides/...' This is hyperbole. It might sell newspapers, but is not an argument unless supported by data.
- 'The principal problem for refugees is absence of income. Students have no income, therefore students are refugees'. This is a very poor argument and untrue in the majority of cases. Some students are refugees, most are not.
- Going OTT with language or throwing in jargon to impress the reader. This is a typical journalistic device, involving an emotional rather than factual appeal to the reader. 'Inner-city housing in Thriftston was run down, the area generally subject to neglect and depopulation' is a more considered academic statement than, 'Thriftston: an abandoned, riot-torn, rubble-strewn, neglected and deserted city', unless of course the situation is extreme. Describing Kabul in 2012 in these terms would be questionable; for central Sheffield it would be seriously biased.
- 'It is unquestionably clear that bacteria control all soil processes'. Words which sound very strong like *clearly, manifestly, undoubtedly, all, naturally* and *obviously*, will influence the reader into thinking the rest of the statement must be true. Overuse of strong words is unhelpful, the written equivalent of browbeating or shouting; used sparingly strong words have greater effect.
- 'The Gulf of Mexico is dead' or 'All penguins are found in the Antarctic'. The global scope of geographical and environmental issues means that an exception can be found to almost every generalization.

Inductive or deductive? Which approach?

In presenting an argument, orally or on paper, give some thought to the ordering of ideas.

Inductive writing starts with specific evidence and uses it to draw general conclusions and explanations. You must decide what suits your material. In general, use the inductive approach when you want to draw a conclusion. (Present a series of facts and case examples, and draw conclusions.) If you write deductively, you begin with the general idea and then follow on with examples.

The **deductive** approach is useful when you want to understand cause and effect, test an hypothesis or solve a problem. (Explain the general position; outline a series of arguments where A causes B, x changes y.)

An example of argument styles and good practice can be found in Holloway and Valentine (2001:130), where their inductive example, is 'pulling a rabbit out of the hat' style, whereas the deductive approach is their 'setting out your stall' style.

In developing arguments seek supporting case studies in soils, geology, agriculture, politics, medical, engineering, economics and other discipline journals as appropriate for your study, not just the GEES literature.

> If you are unsure what someone means, ask for clarification. All feedback on your ideas is brilliant, and needs to be thought about. Take it seriously, think it though, incorporate or reject.

Good and bad arguments

Watch out for illogical arguments creeping into your work. The following examples are rather obvious. Be critical in reviewing your own work: are there more subtle logic problems? It is easy to spot logical errors when sentences are adjacent; less easy when there are over a thousand words in between.

☹ **Circular reasoning** – check that conclusions are not just a restatement of your original premise. E.g. 'Urbanization is a continuous process. We know this because we can see building going on around us all the time'.

☹ **Cause and effect** – be certain that the cause really is driving the effect. E.g. 'Students are forced to live in crime-ridden areas, crime becomes part of the student way of life'. and 'Most faults are observed in quarries, showing that quarrying causes faulting'.

☹ **Leaping to conclusions** – the conclusion may be right, but steps in the argument are missed (see unpacking arguments above). Some arguments are simply wrong! E.g. 'Our survey of five river reaches showed sediment grain size ranged from 0.0002–10 mm in diameter. We therefore concluded that boulders were not a feature of this river'. Did the survey select sediments of all sizes, and how representative of the whole river are the five reaches?

Be clear about the difference between arguments where the supporting material provides clear, strong evidence and those where there is statistical or experimental evidence supporting the case, but uncertainty remains. In student projects and dissertations, time usually limits experimental and fieldwork: 'There should be 150 samples, but, tragically, there was only time to get six'. Be clear about the limitations. Statements like, 'On the basis of the six samples analysed we can suggest that...' or 'The statistical evidence suggests ... However the inferences that may be drawn are limited because sampling occurred over one summer and at three sites in Kowloon' are very acceptable. Qualifying statements of this type have the additional merit of implying that you have thought about the limits and drawbacks inherent in your research results (*more marks*). Watch out for arguments where the author gives a true premise but the conclusion is dodgy. Just because you agree with the first part of a sentence does not mean that the second part is also right. Keep thinking right through to the end of the sentence.

Having articulated an argument, do you buy it? Why? Why not? What is your view? Look at **Try This 11.1** as a starter.

 **TRY THIS 11.1 – Logical arguments**

Consider the following statements. What questions do they raise? Are they true and logical? What arguments could be amassed to support or refute them? See p 298 for some responses.

1 Divergent plate margins occur where two plates are moving away from each other, causing sea-floor spreading.

2 Rainforests store more carbon in their plant tissue than any other vegetation type. Burning forests release this stored carbon into the atmosphere as CO_2. The net result is increased CO_2 in the atmosphere.

3 If the system produces a net financial gain, then the management regime is successful and the development economically viable.

4 Urban management in the nineteenth century aimed to reduce chaos in the streets through paving, lighting, refuse removal and drainage, and improving law and order.

Language that persuades

Strong words often connote weak arguments.

If you use strong statements like 'Clearly we have demonstrated...', 'This essay has proved...' or 'Unquestionably the evidence has shown ...', be sure that what you have written really justifies the hype. It is worth thinking about your stock of linking phrases. Use **Try This 11.2** to review what you currently use, and then look at the answers to see if there are others to add.

TOP TIPS

→ When reviewing your work, read the introduction and then the concluding paragraph. Are they logically linked?

→ Keep the brain engaged.

TRY THIS 11.2 – Logical linking phrases

Identify words and phrases that authors use to link arguments by scanning through whatever you are reading at present. Some examples are given on p 298.

11.5 Opinions and facts

It is important in reading and writing to make a clear distinction between **facts** supported by evidence, and **opinions** which may or may not be supported. Telling them apart takes practice. Think about the evidence offered and what else might have been said. Has the author omitted counter-arguments? Use **Try This 11.3** to help distinguish the two.

TRY THIS 11.3 – Supported facts?

Select a couple of newspapers and grab a pen.

- Read an article, underlining the sections that are supported facts. Identify the arguments made. Circle the opinion or commentary section.
- Read about six articles of different lengths and styles. Then look at the balance of opinion and facts, how do different types of writing change the balance of evidence?
- Then play the same game with a couple of journal papers you need to read anyway. You will see how different sections are structured with evidence, and the author's conclusions.

11.6 Unfashionable arguments

Examining all sides of a question can be particularly difficult if moral, unfashionably moral or ethical elements are involved. Consider how you would discuss the causes of urban rioting: you might say that people riot because it is fun; a group reaction; an opportunity to take personal revenge; a way of livening up a dull, hot evening; an overreaction to a minor misunderstanding that escalated beyond control and reason. You could talk about the 'fact' that some people are 'evil', 'not made evil by circumstance', and state that behaviour is not solely determined by upbringing, school, affluence and employment prospects. A more PC (politically correct) or 'right-on' answer might ignore these elements and discuss social deprivation, unemployment, police brutality, substandard housing, (mis)use of drugs and racial tension. The rioters may be portrayed as victims of circumstance, 'it's not their fault', rather than the active participants, inciters, throwers of bricks and bottles; people responsible for their own actions.

These issues are, of course, enormously difficult. Interlocked, interlinked and coloured in personal understanding by background and culture. The challenge for you, as an unbiased (is this possible?) reporter, is to strive to see all perspectives, including the 'moral' and the unfashionable. If the issue is 'debt', is it because 'credit' is too easily given, personally or nationally? If shoplifting is an economic problem, is it because 'shops make goods too accessible to the customer'?

One entry point might be to think of the headlines that will never make the papers:

NO RIOTING IN LOS ANGELES/TOXTETH/BRISTOL/PRETORIA ON 36,518 DAYS LAST CENTURY

Why were there no riots in most cities on most nights?

 TOP TIPS

As you read any journal article highlight:

→ the arguments,

→ the statements that support the argument, and

→ the statements that counteract the main arguments.

Ask: Is the information balanced?

11.7 Examples of arguments

It is impractical in this text to reproduce an essay and discuss the quality of the arguments. Rather, some students' answers, together with some examiners' feedback, are used to make points about the adequacy of the arguments. This section could be useful when revising for short-answer examination questions.

As an exercise in exemplified brevity, expressing ideas or concepts in two or three sentences, or a maximum of 50 words, is a good game. Consider these three answers to the question:

1 Define feedback. (Two sentences maximum.)

   A *Feedback occurs where part or all of the output from one process is also the input for another process. This can be positive, negative or both.* (26 words)

   B *This is where the system effectively alters itself. Positive feedback is where the signal is reinforced; the converse is called negative feedback.* (22 words)

   C *Feedback is the action by which an output signal from a process is coupled with an input signal (Figure 11:3). Feedback may be positive, which reinforces any changes, e.g. the greenhouse effect or change in albedo, or negative, which ameliorates any changes.* (52 words, including the diagram and its caption.)

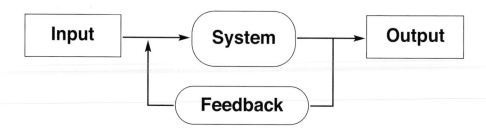

Figure 11:3 A simple feedback loop

These are not perfect answers, but the third statement has a little more technical content, examples, and the diagram reinforces the argument. This, and the next question, asks for a supported, factual response rather than an argument.

2 What is albedo?

A *Albedo is the ratio of the total incident reflected electromagnetic energy to the total incident electromagnetic energy falling on a body. Albedo, expressed either as a % or on a 0–1 scale, varies between surfaces, and generally increases with roughness. The cryosphere has a high albedo, 70–80%, whilst the tropical rainforest has a relatively low value of 20–30%. (61 words)*

B *Albedo = $\dfrac{\text{Total radiation reflected}}{\text{Total Incoming radiation.}}$*

It is measured on a 0–1 scale where 0=low, 1=high. The average albedo of the earth is 0.3. (30 words)

I think both these definitions would get full marks since both define albedo, but if there are bonus marks going, or within an essay, the first answer has added value. Using the equation format (B) perhaps makes the second answer clearer than the first. Inserting equations in essays is fine. Both answers have indicated the units involved.

These following examples build up an argument with evidence.

3 Explain the steps involved in deduction in science.

A *Deduction is where a relationship is deduced from scientific phenomena, whereas induction looks for relationships within data or information. In deduction, a general idea, such as rivers run down hill, is formulated as a hypothesis that may be tested, as in: 'the direction of flow is determined by gradient'. Having defined the hypothesis, it can then be tested in a scientific manner in order to verify or disprove it. Tests might involve primary data collection through direct observation or analysis of maps, photographs, and records of flow. Following analysis, the hypothesis will be accepted, rejected or accepted with caveats.*

Appropriate deductions might involve a statement like, 'The hypothesis was shown to be generally true for ten rivers in southern India. Further research is required to test whether these results hold true in additional areas'.

The first sentence is a little bold (is it right?), but indicates the writer is aware of an alternative, inductive approach. The remaining sentences have integrated examples. I hope you will agree that this answer gets more marks than the following bullet-style answer:

B *The steps in deductive reasoning are: Have an idea; develop a working hypothesis; test the hypothesis experimentally through data collection and analysis; decide whether to accept or reject the hypothesis and further explain how the world works in this case, and then reassess or redefine the problem if the research is to continue.*

This short statement answers the question at a level that will get a pass mark, but it has no discipline content linking the question to examples from geography, earth or environmental science. The first answer will score additional marks if the lecture or main text example was not illustrated with river flow information.

4 Discuss the value of a structured interview compared with a postal questionnaire.

A structured interview has the benefits of being more personal, the interviewer gets a 'hands-on' feel for the data, and problems can be detected and dealt with as the interviews progress, for example, by extending adding or dropping some questions. The interviewee should feel involved, which is likely to encourage full answers and s/he may volunteer additional information. There is a problem with the honesty of answers in all survey research, but a one-to-one interview is more likely to elicit honest responses. The number of interview refusals is likely to be small. Conversely, interviews are very time consuming in both administration and analysis, especially if there are audiotapes to interrogate.

A postal questionnaire can feel very impersonal, there is little motivation to complete, (even Reader's Digest offers wear thin). Since the number of returns may be as low as 15 per cent, it must be sent to a large population to account for non-returns. Some respondents may misunderstand questions, so instructions must be clear and questions unambiguous.

Both approaches require careful preparation of questions, with questions structured in a user-friendly order. A pilot survey should be carried out in each case and the schedule adjusted in response to the results.

If we are looking at argument in isolation then this is good. BUT, this answer would be as useful in a sociology or psychology essay and gain the same marks. It has argument, structure and lots of good points, but no links to geography or environment. There are no GEES examples and no references to the many authors who have written about social survey techniques. It needs the discipline examples and understanding to do better than 45–50 per cent.

 TOP TIPS

→ Avoid 'cop-out' statements like, *'Others disagree that ...'* Who are these others? There are marks for *'Knowitall (2010) and Jumptoit (2010) disagree, making the point that ...'*

→ No examiner gives marks for the use of *'etc'*.

Yes, but ...

In addition to logical linking phrases, keep a list of caveat or 'yes, but' statements handy. There are alternatives to 'however', like: *consequently; as a result; by contrast; thus; albeit; therefore; so; hence; nonetheless; despite the fact that; although it has been shown that.* More extended versions include:

> *A rigorous qualitative geographer might argue that the results of this research are ethereal, confused and disorganised, and that some structure would have helped the project; The author gives an interesting but superficial account of ..; The figures show ... but were not able to support or refute the main hypothesis because ...; The succeeding tests support this criticism because ...; The outcome may be influenced by ...; Only a few of the conclusions are substantiated by the experimental analyses.*

Build on this list as you read. As you research, use these 'yes, but' phrases to focus thinking and to draw valid and reasoned inferences. Be critical (within reason) of your writing and thinking. This means allocating time to read critically, remove clichés and jargon and add caveats and additional evidence.

Argument Checklist

Use this when reviewing the early drafts of a report or essay. Have I got ...

✓ Short clear sentences, with references?

✓ A deductive or inductive approach throughout?

✓ The main, supporting and counter arguments in different paragraphs?

✓ A good balance of supporting and refuting evidence for each argument? Use Figure 11:1.

✓ Case examples as evidence, a balance of GEES and non-GEES examples?

✓ High level of technical content (equations/diagrams/maps/references)?

✓ The points I want to make?

✓ The appropriate style and detail for the intended audience?

11.8 References

Bonnett A 2008 *How to Argue: Essential skills for writing and arguing convincingly*, Pearson Education, London

Hay I and Bass D 2002 Making News in Geography and Environmental Management, *Journal of Geography in Higher Education*, Directions, 26, 1, 129–142

Holloway SL and Valentine G 2001 Making an Argument: writing up human geography projects, *Journal of Geography in Higher Education*, Directions, 25, 1, 127–132

Pirie M 2007 *How to Win Every Argument: The Use and Abuse of Logic*, Continuum International Publishing Group, London

Russell S 2001 *Grammar, Structure and Style*, 3rd Edn., Oxford University Press, Oxford

Wordsearch 2

Find the 30 countries hidden in the wordsearch below. Answers on p 299.

```
D  R  V  T  A  N  N  D  R  A  I  L  U  T  S
I  J  Z  I  M  B  A  B  W  E  A  L  P  E  E
T  N  I  D  A  I  I  A  T  M  G  Y  A  R  N
W  A  A  B  R  L  A  D  A  U  G  I  I  M  E
A  H  B  E  O  I  A  D  N  E  N  T  N  A  G
C  I  G  O  B  U  A  L  C  U  R  I  I  H  A
I  I  B  I  T  G  T  A  G  E  R  L  S  G  L
N  W  M  M  A  S  M  I  A  E  A  U  U  I  Z
O  A  A  S  A  E  W  C  B  M  R  I  B  A  A
N  H  C  L  R  G  O  A  O  E  N  I  A  L  M
T  A  T  O  A  N  R  S  N  E  A  R  A  O  B
R  N  O  O  G  M  C  N  A  A  A  E  Y  G  I
B  N  N  O  S  L  I  B  E  R  I  A  B  N  A
N  O  B  A  G  E  R  W  A  N  D  A  I  A  R
A  Y  N  E  K  H  L  N  A  D  U  S  L  W  D
```

12

Listening

Hearing is easy, listening is tough.

GEES degrees with their mix of lectures, seminars and discussion groups can involve 12 or more hours of listening and note-making each week, so listening skills are important. People listen all the time, but don't always mentally process the information. Various studies suggest people remember less than 50 per cent of what is said to them, and in a lecture if you are distracted by making notes this can be reduced further. While much listening is an automatic activity, it is increasingly cited by employers as a vital skill for effective business performance. Being a good listener who hears and understands most of what is said will potentially improve your relationships with people, avoid misunderstandings, and in business will give the impression of valuing the customer or client, and of taking the time to understand what the speaker is saying.

Interviewing is a standard qualitative research methodology in geography and environmental studies. Therefore, being good at listening has the potential to enhance the quality of your research understanding and inferences.

Listening is not the same as hearing, it is a more active and interactive process. Listening involves being ready to absorb information, paying attention to details, and the capacity to catalogue and interpret the information. In addition to the actual material and support cues, like PowerPoint slides, there is information in the speaker's tone of voice and body language. As with reading, the more knowledge you have, the more you are likely to understand, which means that listening is a skill involving some preparation, a bit like parachuting.

12.1 Listening in lectures

Arriving at a lecture with information about last night's activities or juicy scandal is normal, but the brain is not prepared for advanced information on radiolarian ooze or medieval housing. Some lecturers understand that the average student audience needs five minutes' background

I am a bit dyslexic, not enough to get extra time, so being able to read the slides slowly, before the lectures, made such a brilliant difference to understanding in the lectures. I was able to listen to what was said.

briefing to get the majority of brains engaged and on track. Others leap in with vital information in the first five minutes because 'everyone is fresh'! Whatever the lecture style, but especially with the latter, you will get more from the session having thought 'I know this will be an interesting lecture about ...' and scanned notes from the last session or library. Assuming from the start that a lecture will be dull usually ensures that it will seem dull.

TOP TIPS

→ Where the PowerPoint slides for your modules are on the VLE ahead of the lecture, look at them before going to the session. They won't make perfect sense, but you will have an idea about what is coming up and will remember much more.

→ A lecturer's words, no matter how wise, enter your short-term memory. Unless you play around with them and process the information into ideas, making personal connections, the words will drop out of short-term memory into a black hole. Think about the content and implications as the lecture progresses.

→ You may feel a lecturer is wildly off beam, making statements you disagree with, but do not decide he or she is automatically wrong – check it out. There might be dissertation possibilities.

→ Keep a record of a speaker's main points. Make notes.

→ Be prepared for the unpredictable. Some speakers indicate what they intend to cover in a lecture, others whiz off in different directions. This unpredictability can keep you alert, but if you get thoroughly lost, then ask a clarifying question (mentally or actually), rather than 'dropping out' for the rest of the session.

→ If you feel your brain drifting off, ask questions like: 'What is s/he trying to say?' and 'Where does this fit with what I know?'

→ Have another look at p 5, expectations of lectures.

→ Treat listening as a challenging mental task.

12.2 Listening in discussions

Discussion is the time to harvest the ideas of others. With most topics such as ecotourism, contaminated land management and political geography there is such a diversity of points of view that open discussion is vital. Endeavour to be open-minded in looking for and evaluating statements which may express very different views and beliefs from your own. Because ideas fly around fast, make sure you note the main points and supporting evidence (arguments) where possible. Collating notes and adding references after a discussion is crucial. This involves ordering thoughts and checking arguments that support or confute the points. Have a look at **Try This 12.1** and think about how you score on effective listening.

> *I can text, talk and listen to my iPod at the same time, just don't ask me about what is going on!*

TRY THIS 12.1 – Assess your listening effectiveness in discussions

Think back over a tutorial or a recent conversation. Rate and comment on your input for each of these points. Alternatively, score this for a friend, and then think about your skills as listeners.

Score your effectiveness on a 1–4 scale, where 1 is No, and 4 is Yes		Reflections Thoughts
Did you feel relaxed and comfortable?	1 2 3 4	
Did you make eye contact with speaker?	1 2 3 4	
Were you making notes of main points and personal thoughts during the discussion?	1 2 3 4	
Did you discuss the issues?	1 2 3 4	
Were you thinking about what to say next while the other person was still talking?	1 2 3 4	
Did you ask a question?	1 2 3 4	
Did you get a fair share of the speaking time?	1 2 3 4	
Did you empathize with the speaker?	1 2 3 4	
Did you accept what the others said without comment?	1 2 3 4	
Did you interrupt people before they had finished talking?	1 2 3 4	
Did you drift off into daydreams because you were sure you knew what the speaker was going to say?	1 2 3 4	
Did the speaker's mannerisms distract you?	1 2 3 4	
Were you distracted by what was going on around you?	1 2 3 4	

12.3 Telephone listening

Some research interviews must be conducted over the telephone, and companies are increasingly conducting first job interviews on the telephone. Listening and talking under these circumstances are difficult to do well. If Skype or video-conferencing facilities are available you can pick up on facial and body-language clues, but these are missing in a telephone interview. This skill improves with practice. Pilot interviews are vital: practise with a friend.

 TOP TIPS

Telephone interviews

Always. Every time:

→ Find a quiet room to phone from and get rid of all distractions (radio, mobile, TV, dog ...).

→ Be prepared by laying your notes out around you and have at least two pens, and spare batteries if using a recorder.

Research interviews

→ Plan the call in advance. Organize your questions and comments so you can really concentrate on the responses and implications.

→ Make notes of main points rather than every word, and leave time after the call to annotate and order the responses while the information is fresh in your brain.

→ Query anything you are unsure about. Be certain you understand the interviewee's nuances.

→ Show you are listening and interested without interrupting, using 'yes', 'mmm', 'OK' and 'great'.

→ Search for verbal clues, like a changed tone of voice, to 'hear between the lines'.

→ Don't think you know all the answers already. If you disengage, the interviewee will become less engaged, less enthused and be a less productive informant.

→ Curb your desire to jump in and fill pauses; let the speaker do most of the talking. Silences are OK.

Job interviews on the telephone

→ Prepare in advance as you would for a visit to a company, by researching the company background and position.

→ Make notes of points as you speak and query anything you don't understand.

→ Be enthusiastic! All GEES graduates have lots to offer.

> *I found it helped to dress like I was going to a real interview, made me concentrate ... Odd at home, but no one knows.*

→ Be formal in your conversation – there is a job in prospect. It is easy to drop into a colloquial, conversational mode as if chatting to a friend, which you would not do in an office interview.

12.4 One-to-one interviews

Most of the tips for telephone interviews apply equally to personal interviews, whether for research or with a potential employer. If you are the interviewer, choose locations where you will not be interrupted. Take a coffee break in a long session to give both your own and your interviewee's brain a break. Remember, people can speak at about 125 words a minute, but you can listen and process words at 375 to 500 words a minute, so it is easy to find your brain ambling off in other directions. Wool gathering or star gazing are not good!

Jumping to conclusions in a discussion is dangerous as it leads you to switch off. Possibly the speaker is going in a new direction, diverting to give additional insights. Watch out for those 'yes, but' and 'except where' statements.

> *I have all these notes, what did he say?*

Got a difficult customer? Let them talk. They will feel in charge and get the idea you agree with their discourse, which you may or may not.

→ TOP TIPS

→ Really good listeners encourage a speaker by taking notes, nodding, smiling and looking interested and involved. Verbal feedback (replying) is often better as a statement that confirms what you have heard, rather than a question which will probably be answered by the speaker's next statement anyway. 'Did you mean ...?' or 'Am I right in thinking you are saying ...?' Unhelpful responses include yawning, looking out of the window, writing shopping lists and going to sleep. Relating similar personal experiences or offering solutions does not always help as, although you think you are offering empathy or sympathy, it may appear that you just turn any conversation around to yourself. (See section 12.4.)

→ Remain objective and open-minded. If you are emotionally involved you tend to hear what you want to hear, not what is actually said.

→ Keep focused on what is said. Your mind does have the capacity to listen, think, write and ponder at the same time; there is time to summarize ideas and prepare questions, but it does take practice.

→ Make a real attempt to hear and understand what the other person is saying.

→ Think about what is not being said. What are the implications? Do these gaps need exploration?

→ When you are listening, interruptions need sensitive management. If you answer the phone or speak to the next person, the person you are speaking to will feel they are less important than the person who interrupts. If you do this to someone in business, they are very likely to take their business elsewhere. So turn the phone off and shut the door.

No one listens until you make a mistake.

12.5 Further resources

There are a variety of texts and online resources about listening. Search using Active Listening, Critical Listening and Listening Skills as keywords. Some of the literature relates to counselling where listening is particularly important, you may find the ideas presented are interesting and useful.

13

Oral presentations

Tension's what you pay when someone's talking.

GEES degrees are littered with speaking opportunities. This is a really good thing! Somewhere on your CV you can add a line like, 'During my university career I have given 25 presentations to audiences ranging from 5–165 using PowerPoint with embedded video and podcasts'. This impresses employers. Many GEES graduates forget to tell employers that they have had these opportunities. Presenting helps you to manage your nerves and allows you to become familiar with the question-and-answer session that follows. Grab all opportunities to practise your presentation skills. They all count for CVs, from five-minute presentations in tutorials, to seminars and mini lectures.

This chapter has useful advice, but presentations are essentially a visual performance. Think about your lecturers. What do you want to avoid doing? What would be good to do? Check out YouTube and flickr: How to give ... funny, knockout, effective presentations, and public speaking.

You need a well-argued and supported (*add references*) message to enrapture (ideally) the audience. Other chapters explain how to get the information together; this one is about practical presentation skills. The four most important tips are:

1 Suit the style and technical content of your talk to the skills and interests of the audience. Make the content accessible, so that the audience wants to listen and ask questions.
2 Buzan and Buzan (1993) show that people are most likely to remember items:
   (a) from the start of a learning activity
   (b) from the end of a learning activity
   (c) associated with things or patterns already learned
   (d) that are emphasized or highlighted as unique or unusual
   (e) of personal interest to the learner.

Tailor your presentation accordingly. Help the audience to think and understand.

3 Get the message straight in your head, organized and ready to flow. Lack of confidence in the content → insecure speaker → inattentive audience → bad presentation = low marks.
4 Remember, the audience only gets one chance to hear you. Keep a short

presentation short, essential points only. Make clear links between points and add brief, but strong, supporting evidence (*references are good*).

You must have a plan, and stick to it on the day. Basically it is down to PBIGBEM (Put Brain In Gear Before Engaging Mouth).

13.1 Style tips

- **Can I read it?** Reading a script will bore both you and the audience. Really useful tutors (you will hate him/her at the time) will remove detailed notes, asking you to 'tell the story in your own words'. You are allowed bullet points on cards to remind you of the main points, but that is all. Illustrate your talk with maps, equations and diagrams, but remember that reading PowerPoint slides aloud is another cop-out.
- **Language.** A formal presentation requires formal speaking. Minimize colloquial language, acronyms and paraphrasing, and limit verbal mannerisms like the excessive use of *hmm, umm, err, neat, cool*, and *I mean*. Help the audience by putting new words or acronyms on a handout or slide.
- **Look into their eyes!** Look at the audience and smile at them. If they feel you are enthusiastic and involved with the material, they will be more involved and interested.
- **How fast?** Not too fast, slower than normal speech, because people taking notes need time to absorb your ideas, to get them into their brains and onto paper. You know it, but it is new for your audience. Watch the audience to check if you are going too fast, or they are falling asleep.
- **Stand or sit?** Remember that sitting encourages the audience to feel it is a less formal situation and one where it is easier to chip in with comments.
- **How loud?** So that the audience can hear, but do not feel you are shouting. Ask someone you trust to sit at the back and wave if you are too loud or too quiet – getting it right takes practice. Tape record yourself sometime. When you have finished laughing at the result, have a think about whether you speak at the same pace and pitch all the time. Changing pace and pitch, getting excited about the material and showing enthusiasm help to keep your audience attentive and involved.

  All slides must be visible from the back of the room.
- **Repeating points?** You can emphasize points by repetition, or by simultaneously putting them on a slide or flip chart, but only repeat the important points.

13.2 Crutches (slides, flip charts, video …)

Audiences need to understand your message. Visual assistance can include some or all of:

☺ A title slide that includes your name and email address, so the audience knows who you are and where to find you!

☺ A brief outline of the talk: bullet points are ideal, but adding pictures is fun – just make sure the message is clear. Is Figure 13:1 clear enough or OTT?
(For a 10-minute presentation it is too ambitious. There are too many pictures. The punctuation is muddled. Capitals must to be used consistently.)

☺ A map – the location (geography) matters – tell the audience where you are talking about.

☺ Colours on complex diagrams help to disentangle the story.

☺ Graphs, cross-sections, and pictures.

☺ Finally, a summary sheet. This may be the second outline slide shown again at the end, or a list of the points you want the audience to remember.

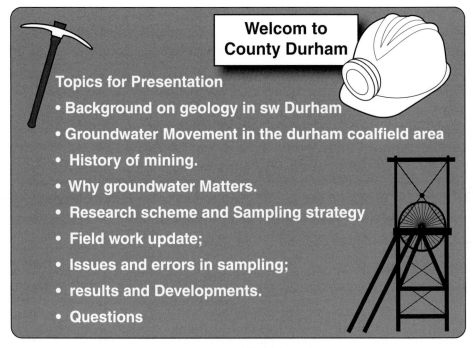

13:1 Slide 2 – Setting the scene for a 10-minute presentation

PowerPoint Slides

Preparing effective slides takes practice. What does not work: small writing, too many colours, words from edge to edge. Misuse of or inconsistent Use OF capitals Doesn't help Either.

Good practice includes:
- Slides prepared carefully in advance.
- SPELL CHECK EVERYTHING.
- Minimum font size 24. Keep writing VERY BIG and messages short!
- Adding video, cartoons, clipart items or photographs makes the message more memorable, they really help with very technical or dull messages. Make sure visual items are totally relevant.
- Use diagrams and maps from books and papers, scan them BUT you must cite their source.
- Some audience members are colour blind. Avoid green and red together, black with blue, and black on red. Yellow and orange do not show up well in large lecture theatres, and orange and brown are not easy to distinguish from 30 rows back.
- If you have lots of information and time is short, give the audience a handout with the detailed material and use the talk to summarize the main points.
- Reference every source, this especially includes photographs (Figure 13:2).

Before giving a talk, investigate the room, computer, projector and microphones. Find the on/off switches, the plugs, light switches, and the passwords in case the computer crashes and needs rebooting. 'Sod's law of presentations' says the previous speaker will breeze in, give a brilliant talk and turn everything off as s/he finishes leaving you to reset the computer while the audience watches you! KEEP COOL and DON'T PANIC.

Run the slides through in advance to ensure slides show OK, can be read easily and that video and internet links are working.

Figure 1 Athabasca glacier, British Columbia

Figure 1 Athabasca glacier, British Columbia (Personal photo, Kneale 2009)

Figure 13.2 Unacceptable and acceptable PowerPoint slides

Flip charts

Write big and make sure your pens are full of ink. Left- and right-handed persons need the chart in slightly different positions. Check beforehand what the audience can see and adjust your position accordingly. If spelling is a problem, flip charts are a BAD IDEA. The audience will remember your spelling errors rather than your message.

13.3 What to avoid

☹ **Getting uptight.** All speakers are nervous. THIS IS NORMAL. Take deep breaths and relax beforehand. If you are well prepared there is the time to walk slowly to the room and have a coffee beforehand. Get to the session early enough to find the loos, lights and seats. Ensure that the IT, flip chart, handouts, notes and a pen are in the right place for you and the audience. Take a lot of deep breaths. YOU WILL BE FINE.

☹ **Showing the audience you are nervous.** Ask your mates to tell you what you do when speaking. Everyone gets uptight, but fiddling with hair, pockets, clothes, keys, pencil, ears and fingers distracts the audience. They watch the mannerisms and remember very little of the talk. Practise speaking.

☹ **Overrunning.** They reckon you can speak at about 100–120 words per minute. Practise in advance with a stopwatch (there is one on most cookers), reducing the time by 10 per cent.

13.4 Handling sticky situations

☺ Ignore late arrivals, unless they apologize to you, in which case, smile your thanks. Going back interrupts your flow, irritates the people who arrived on time, and encourages people to come late next time.

☺ Time is up and you have 20 more things to say. This is bad planning and usually only happens when you have not practised the material aloud. You can read your material much, MUCH faster than you can speak it. However, if you do run over time either skip straight to your concluding phrases OR list the headings you have yet to cover and do a one-line conclusion.

☺ Good question, no idea of the answer. Say so. A phrase like 'That's a brilliant thought, not occurred to me at all, does anyone else know?' will cheer up the questioner and hopefully get others talking. To give yourself some think time say: 'I am glad this has come up ... I wanted to follow up that idea but couldn't find anything in the library, can anyone help here?' and 'Great question, no idea, how do we find out?'

☺ The foot–mouth interface. They happen. GEES favourites include '… improved the engineering of buildings so they stood a better chance of collapsing in an earthquake'; 'The gravity model can be heavy going to program'; 'What we found was that the desert was really, really dry, quite arid in fact'.

☺ Make notes of feedback points in the Q&A sessions.

 TOP TIPS

→ Self-assess your bathroom rehearsals against your department's presentation feedback form, or use Figure 13:3 to polish a performance.

→ Watch your language. Try not to use colloquial speech or to substitute less than technical terms, 'The results seemed a bit unlikely', 'There were loads of rocks …' or 'Clouds, mmm, yes, they are fluffy ones'. Strive in presentations to use a formal, technically rich, academic style.

→ Practise your talk on the parrot, the bathroom wall and … Speaking the words aloud will make you feel happier when going for gold. It gives you a chance to pronounce unfamiliar words, like equifinality, fluorapatite or *Rhytidiadelphus squarrosus*, and get them wrong in the privacy of your own bus stop. If in doubt about pronunciation ask someone, the librarian, your mum, anyone.

→ Most marks are given for content, so research, research and add references.

He (tutor) made us include the references as we presented and discussed. That was really strange… makes you remember who wrote the papers, so good revision.

I hated presentations, I don't like the sound of my voice. We had to do a short presentation at every tutorial. It was a real pain to prepare and my tutor made us repeat points so we used more complicated words. But we did get more practise than any of the other groups so by the final year our group was really good. Useful now in my job. (Graduate 2009)

Oral Presentation		Great	Middling	Oh Dear!	
Speed	Spot on				Too fast/slow
Audibility	Clear and distinct				Indistinct
Holding attention	Engaged audience				Everyone asleep
Organization	Organized logical structure				Scrappy and disorganized points
Relevance	All material on topic				Random and off the point
Academic support	References for factual and visual materials				No references
Slide production	High-quality clear and referenced				Unclear and unreferenced
Handling questions	Thoughtful answers				Limited ability to extend the discussion
Teamwork	Good balance of team input and questions				Unequal and unbalanced presentation

Figure 13.3 Self-assess an oral presentation. Use when your department version isn't available

13.5 References and resources

Buzan T and Buzan B 1993 *The Mind Map Book*, BBC Books, London

Hay I 2006 *Communicating in geography and the environmental sciences*, 3rd Edn., Oxford University Press, Melbourne

Look critically at real presentations on YouTube. Spot the good and not so good points.

Add up 2

Work out the totals for the top of the pyramids. Answers on p 299.

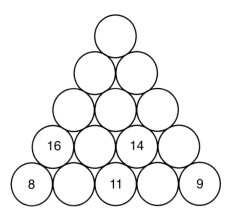

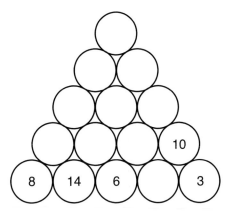

14

Discussion

Why is it that when you discuss Murphy's Law,
something always goes wrong?

Research thinking happens during discussions in GEES workshops, tutorials and seminars. In most jobs, being able to discuss topics calmly, fairly and professionally is essential. Talking through the details of a topic leads to a greater degree of understanding and learning. Consequently modules have lots of discussion opportunities, places where ideas are created and innovated. Discussions vary from very short, for example in a lecture where you discuss something with a neighbour, to full scale seminars with a five-minute scene-setting presentation that starts a 40-minute discussion.

To learn effectively from discussions a relaxed atmosphere is vital, letting people think about the content and make notes about what others are saying. Making notes is crucial to capture ideas as they whiz around, and helps you to sort out what you want to say. Ineffective discussions occur when people are worrying about what to say next; mentally rehearsing what to say rather than listening to the person speaking. Some of this is nerves, which will calm down with practice, but in the meantime preparing fully is the best way to lower your stress levels. You have background and specific information to share.

Be positive about seeking the views of others and value their contributions. Employ open-ended questions, those which encourage an elaborated, rather than a brief yes or no answer. 'What are the main resort facilities of Penang?' is an open-ended question, more useful than 'Do you like Penang?' 'What do you think about green issues/global warming/transport in New Delhi?' are good questions, but 'What do you think ...' is a bit general. Use phrases like 'What are the advantages of global warming?' or 'Do you consider the Internet's global, unrestricted basis

> *What you are really saying is you want us to discuss like we would write in an essay?*

> YES, with technical terms and references. That's what makes a strong academic argument, oral or written.

will influence the actions and attitudes of individuals?' or 'What should be valued about the lake and its environs as a community recreation resource?'

Keep up the quality of argument in discussion (*include references*). For example, discussing spatial variability in river sediment transport, you might make a general point: 'There are natural variations in sediment loads and fluxes with time and changing environmental conditions'. This general argument would be strengthened, getting more marks, by adding references and examples as you speak. You might say: 'Kettner *et al.* in their 2010 paper report a simulation of fluvial sediment fluxes in a river basin in Colombia. They show the sediment discharge is influenced by the sub-basin geology, climate variations and human impact. Deforestation and mining make particular reference to the fluxes. Similarly the paper by McCarney-Castle *et al.* (2010), looking at fluxes from five US rivers, ...' This has added additional technical information, demonstrated your research skills and given more than one case example.

14.1 Types of Discussion

Brainstorming

Brainstorming is a great way of collecting a range of ideas and opinions and getting a group talking.

The process involves everyone calling out points and ideas. Someone keeps a list on a whiteboard or flip chart so everyone can see. A typical list has no organization; there is overlap, repetition and a mix of facts and opinions. The art of brainstorming is to assemble ideas, including the wild and wacky, so that many avenues are explored. The points are

It seemed very ... uncool to make notes about what your mates say ... swotty. I should have started doing it sooner. I missed loads of good stuff at the start ...

reordered and arguments developed through discussion, so that by the end of a session there is a pooled, ordered and critically discussed list. Use **Try This 14.1** as an example of working with a brainstormed list, and use **Try This 14.2** with your next essay.

TRY THIS 14.1 – Working with a brainstormed list

Brainstorming produces a list of ideas with minimal detail and no evaluation. This list was compiled in five minutes during a coastal field trip. Items overlap and repeat, facts and opinions are mixed, and there is no order. Take five minutes to categorize the items. A possible sort is given on p 299.

Physical and human influences on the coastal landscape at _____:

Concreting the slopes prevents soil erosion.

Sea wall prevents undercutting.

Rocks at cliff foot dissipate wave energy.

Groynes prevent longshore drift.

Fine and coarse netting, geotextiles, used on clay slope to help stabilize the soil.

Bolts into the rock face increase stability.

Tourist facilities destroy foreshore view.

Rock armour installed at the foot of recent landslip.

Movement of handrails.

Vegetation eroded from wetter clay slopes.

Tennis courts on old levelled slip.

Complex geology, land slips, big fault structures.

Undercutting of cliffs by waves.

Evidence that paths across the slips are regularly re-laid, cracked tarmac, fences reset.

Café for tourists and visitors at foot of slip area.

Soil overlying clay.

Wall to prevent boulders from cliff hitting the road.

Tree planting on recent and older slip hitting sites.

Drainage holes in retaining walls to reduce soil water pressure.

Vegetation worn on paths and alongside paths.

Car parking demands, unsightly in long views.

Promenade held up by an assortment of stilts, walls and cantilever structures where slope below has been undercut.

Litter needs collecting from beach.

Road built over old stream, presumed piped underground.

Castle fortification, some walls now undermined and eroded.

Hotel, amusement arcade and tourist shops – noisy and unsightly.

Harbour has mix of marina, fishing and tourist fairground rides.

Path covered by slumping clay.

Repairs to sea wall.

Art deco swimming pool, in need of repair.

Car pollution.

Car parking unscenic.

Caravan park on cliff.

Varied quality of coastal path, not always well signposted.

TRY THIS 14.2 – Brainstorm an essay/report plan
Use your next writing task as the focus, brainstorm a list of ideas with colleagues and friends, including references and authors, and use this discussion as the basis for essay planning. Brainstorming 'what I know already' at the start of planning indicates where further research is required.

Role-play exercises

Role-play is used in university modules to help you to understand the multiple dimensions of many issues and different parts to complex jobs and processes. And they are so much more fun than reading. GEES role plays can involve running a community consultation meeting, running a local government consultation meeting, bidding for funds, presenting a project proposal to a board, presenting options for decisions by a planning authority, a press conference …

The planning game was great fun, because we all had different roles and different points to make. It made the arguments much more real, some people got really excited, especially the protesters group. It made me think of that how people who live there would look at our designs.

A good role-play will involve preparing the academic arguments for or against (a development in a national park, drilling a borehole, starting a business) and presenting them in a role (consultant, engineer, lobbyist, pressure group representative, councillor, MP), not necessarily a role you would agree with personally.

Debate

The normal format for a debate involves a contentious issue presented as a motion. It is traditionally put in the form: 'This house believes that …' One side proposes the motion and the other side opposes it. The proposer gives a speech in favour, followed by the rebuttal presenter speaking against the motion. These speeches are 'seconded' by two further speeches for and against, although for reasons of time these may be dispensed within one-hour debates. The motion is then thrown open so everyone can contribute. The proposer and the rebuttal presenter make closing speeches, in which they can answer points made during the debate, followed by a vote. Since GEES issues are rarely clear-cut, a formal debate is a useful way of exploring positions and opinions, and for eliciting reasoned responses (*add references* to your speech).

Oppositional discussion

Oppositional discussion is a less formal version of debate, in which each side tries to persuade an audience that a particular case is right and the other is wrong. You may work in a small group, assembling information from one point of view, and then argue your case with another group which has tackled the same topic from another angle. Remember that your argument needs supporting with evidence, so keep case examples (*references*) handy.

I hated that assessed discussion session, I was totally nervous and didn't sleep before it. It was mega stressful ... as bad as exams... I wasn't happy. It did make me do it so the next session wasn't as bad. I learned quite a lot about preparing for it ... although you think you know what the arguments are and then loads of other things get mentioned.

Consensual discussion

Consensual discussion involves a group of people with a common purpose, pooling their resources to reach an agreement. Demonstrations of good, co-operative discussion skills are rare; most of the models of discussion on TV, radio, and in the press are set up as oppositional rather than consensual. Generally you achieve more through discussing topics in a co-operative spirit and, one of the abilities most sought after by employers of graduates is the ability to solve problems through co-operative teamwork.

Negotiation

Negotiation, coming to an agreement by mutual consent, is a really useful business skill. People practise and improve negotiation skills every day – by persuading a tutor to extend an essay deadline, getting a landlord to repair the heating, by persuading someone else to clean the kitchen. In formal negotiations aim to:
✓ Prepare by considering the issues in their widest context.
✓ Enumerate the strengths and weaknesses of your position. It reduces the chances of being caught out!
✓ Get all the options and alternatives outlined at the start. There are different routes to any solution and everyone needs to understand the choices available.
✓ Check that everyone agrees that no major issue is being overlooked, and that all the information is available to everyone.
✓ Appreciate that there will be more than one point of view, let everyone have their say.
✓ Stick to the issues that are raised and avoid personality-based discussion; s/he may be an idiot, BUT saying so will not promote agreement.

✓ Assuming decision deadlines are flexible, break for coffee or agree to meet later if discussion gets overheated.

✓ At the end, ensure everyone understands what has been decided by circulating a summary.

There are many books on discussion, assertiveness and negotiation skills, do a keyword search to see what your library offers.

⇨ TOP TIPS

→ Being asked to start a discussion is not like being asked to represent your country at football. You are simply 'kicking off'. Make your points clearly and 'pass the ball' promptly. Focus thoughts by putting the main points on a handout or flip chart.

→ Don't wait for a 'big moment' before contributing. Asking questions gets a topic going.

→ Don't be anxious about the quality of your contributions. Say something early on when everyone is nervous and too concerned about his or her own contribution to be critical.

→ Keep discussion points short and simple.

→ Use GEES examples to illustrate and strengthen your argument. Add references from the start.

→ Share the responsibility for keeping the group going.

→ Have a short discussion beforehand to kick ideas about. Meet in the bar, over coffee or supper, somewhere informal. Start talking.

14.2 Discussions online

It is not always possible to get people together for a discussion. Online discussions via email, VLE groups, Skype, Sharepoint or e-conferencing can solve groupwork timetabling problems, and are vital for part-time, year abroad and work placement students. They are great practice for business discussions. One advantage is that you can build research activities into the process. Having started a discussion, you may realize that you need additional information. You can find it and feed it in as the discussion progresses.

Here are some ground rules to help run virtual discussions:

1 Agree a date to finish (two weeks).
2 Agree that everyone must make a minimum number of contributions (three).
3 Agree to read contributions every x days (two or three days).
4 Appoint someone to keep and collate all messages so there is a final record (Joe).
5 Appoint a 'devil's advocate' or 'pot stirrer' to ask awkward questions and chivvy activity (Sandy).
6 Ask someone to summarize and circulate an overview at the end, especially if there is a group report to create (Jenny).
7 Be polite. In a conversation you can see and hear when someone is making a joke or ironic comment. The effect is not always the same on screen.
8 Where further research is required, attempt to share tasks evenly.
9 Replying instantly is generally a good idea; that is what happens in face-to-face discussion, and first thoughts are often best.

> Think flexibly

Some people 'lurk' quietly, listening rather than commenting, which happens in all discussions. Point two, above, should overcome this issue to a certain extent. One of the more off-putting things that can happen in an electronic discussion is someone writing a 3000-word essay and mailing it to the group. This is the equivalent of one person talking continuously for an hour. It puts off the rest of the group as they will feel there is little to add. Try to keep contributions short in the first stages. One good way to start is to ask everyone to brainstorm 4–6 points to one person by the end of day two. These are collated, ordered and mailed around the group as the starting point for discussion and research. (See **Try This** 15.1 – Sharing documents by email.)

> I only read postings of 4 lines or less.

Modules with online discussions will have the process set up, usually via the university VLE. Some students prefer to have discussions on Facebook and other social networking sites as they are more 'private'. Setting up a separate group via Sharepoint or Facebook type sites is fine. What you risk is:

> It was ace having the Skype meetings. Three people in one flat, two in theirs and Jen at home. Worked so much better because we saw each other. ☺

• missing valuable feedback from the lecturer who could be helping the group to think through issues in online discussions, just as she would in a face-to-face discussion, and is doing so with other groups.
• not having the whole class involved. The diversity of ideas will be reduced.

14.3 Group management

The quality of discussion depends, above all, on the dynamics of the group. Some discussions work spontaneously without any problems, others are very sticky. There are no hard and fast rules about behaviour in group discussions, but here are some general points worth considering. Meetings flow well when people take different roles.

☺ Chair (steer) the discussion to keep it to the point, sum up, shut up people who talk too much and bring in people who talk too little.

☺ Keep track of the proceedings.

☺ Inject new ideas.

☺ Are critical of ideas.

☺ Play devil's advocate.

☺ Calm tempers.

☺ Add humour.

Take notes during discussions

Formal meetings have designated individuals for the first two tasks (chair and secretary), but in informal discussions, anyone can take these roles at any time. Everyone is better at one or two particular roles in a discussion; think about the roles you play and about developing other roles using **Try This 14.3**.

⚙ TRY THIS 14.3 – Discussant's role

Sort the roles people take in discussions into those which are positive and promote discussion, and those which are negative (adapted from Rabow *et al.* 1994). How might you handle different approaches? (Suggested answers on p 300.)

Offers factual information.	Gives factual information.
Encourages others to speak.	Helps to summarize the discussion.
Offers opinions.	Asks for examples.
Speaks aggressively.	Is very defensive.
Asks for reactions.	Asks for examples.
Diverts the discussion to other topics.	Is very competitive.
Seeks the sympathy vote.	Keeps quiet.
Gives examples.	Ignores a member's contribution.
Summarizes and moves discussion to next point.	Is very (aggressively) confrontational.
Keeps arguing for the same idea, although the discussion has moved on.	Mucks about.
	Asks for opinions.

14.4 Giving and receiving feedback

Asking tutors is not cheating!

Discussions often involve an element of feedback from and to colleagues and tutors. There are some useful ground rules for feedback. Be constructive because this helps to build everyone's self-confidence, aim to offer options and to encourage the further development of the idea, discussion point or skill. Start with the positive and remember to exemplify your statements. 'I found this a useful and concise summary of the article, the section on moraine types was particularly useful' is much better than 'I liked it'. 'I think it would have helped me to understand it if you could have said a little more about the equity issues' – gives the discussant a way forward and progresses the ideas. Similarly, so do 'I think I understand but could you give an example of ...' and 'I think this is a very different way of looking at the issues to the one I had been considering; this is helping me to see another side to the argument'. Opening up other approaches or options can be very helpful for an individual and group. 'I felt the opening section lacked depth and examples, it didn't really make me feel involved. For me, a stronger start, perhaps with some literature cited would have helped me to jump into your material', or 'This is a really interesting way of tackling ... What would happen if we looked at it from ...' or 'Is it worth exploring other angles? For example...'

Receiving feedback can be awkward, many people are not very skilled at giving it. They may be overcritical, intent on getting their own ideas across and, in some cases, just plain insensitive. This does not make for easy discussions, so try not to feel defensive and 'got at'. Make sure you understand the feedback points you are given, and ask for examples if you think you are getting an off-the-cuff reaction. One useful approach is to note the statements and say, 'Thanks, I will think it over, that's been very helpful'. If, in the cold light of the next day you feel it was unjustified comment, ignore it; if it is helpful advice, use it.

Discuss everything

14.5 Assessing discussions

At the end of each term, or as part of a learning log (Chapter 2), you may be asked to reflect on your contribution to discussion sessions, and in some cases to negotiate a mark for it with your tutor. The attributes an assessor might check for are included in **Try This 14.4**. Assess study buddies, seminars and TV discussants on this basis. What you can learn from those with high scores?

TRY THIS 14.4 – Assessment of discussion skills

Evaluate a discussant's performance on a 1–5 scale, noting what they do well.
1 (useless), 3 (average), 5 (brilliant)

	Feedback on good points
Talks in full sentences.	
Asks clear, relevant questions.	
Describes an event clearly.	
Listens and responds to conversation.	
Discusses and debates constructively.	
Speaks clearly and with expression.	
Selects relevant information from listening.	
Responds to instructions.	
Contributes usefully to discussion.	
Reports events in sequence and detail.	
Is able to see both sides of the question.	
Finds alternative ways of saying the same thing.	
Listens to others, and appreciates their input, efforts and needs.	

Having analysed what makes a good discussant, have a look at **Try This 14.5**.

TRY THIS 14.5 – Self-assessment of discussion skills

Level 1 students at Leeds brainstormed the following list of 'skills they needed to discuss effectively'. Do you have items to add? Select three items where you would like to be more effective and plot a strategy to work on each of these three issues at your next group discussion (e.g. I will not butt in; I will ask at least one question; I will say something and then shut up until at least three other people have spoken).

Being open-minded.	Staying cool.
Listening to both sides of the argument.	Being tolerant.
Using opponents' words against them.	Using good evidence.
Playing the devil's advocate.	Being willing to let others speak.
Thinking before speaking.	Only one person talking at a time.
Summing up every so often.	Being firm.

Some final points

Getting better at discussion and argument needs practice; hearing one's own voice improves one's self-confidence. You can practise in private. Listen to a question on a TV or radio discussion programme. Turn the sound down, take a deep breath to calm down, and use it as thinking time. What is the first point you want to make? Now say it out loud. Subject matter is not important, get in there and have a go. Respond with two points and then a question or observation that throws the topic back to the group or audience. This is a good technique because you share the discussion with the rest of the audience, who can contribute their range of views. Record the programme to compare your answer with the panellists' responses, remembering to look at the style of the answers rather than their technical content.

Where points of view or judgements are needed, seek the opinions of people with different academic, social and cultural backgrounds and experience. Their views may be radically different from your own. Seminars, workshops and tutorial discussions in geography and environmental science are designed to allow you to share these kinds of complementary views. To get the most out of a discussion or conversation:
✓ Be positive.
✓ Ask yourself questions, like 'How will this help me understand ... passenger transport pricing, oscillation ripples, ammonites?'
✓ Make eye contact with the group.
✓ Give other discussants feedback and support.
✓ Aim to be accurate and on the point.
✓ Include examples and references as you speak.

14.6 References and further reading

Annand M 2010 Introduction to Sharepoint, http://it-help.bathspa.ac.uk/onepage_sharepoint.html   Accessed 15 January 2011

Kettner AJ, Restrepo JD and Syvitski JPM 2010 A Spatial Simulation Experiment to Replicate Fluvial Sediment Fluxes within the Magdalena River Basin, Colombia, *Journal of Geology*, 118, 4, 363–379

McCarney-Castle K, Voulgaris G and Kettner AJ 2010 Analysis of Fluvial Suspended Sediment Load Contribution through Anthropocene History to the South Atlantic Bight Coastal Zone, USA, *Journal of Geology*, 118, 4, 399–416

Rabow J, Charness MA, Kipperman J and Radcliffe-Vasile S 1994 *William Fawcett Hill's Learning Through Discussion*, 3rd Edn., Sage Publications Inc., California

University of New South Wales 2008 Discussion Skills for Tutorials & Seminars, Learning Centre UNSW, http://www.lc.unsw.edu.au/onlib/pdf/disc.pdf Accessed 15 January 2011

YouTube has a range of videos with discussions of varying quality. Good for ideas. Search starting points: discussion or debate, with skills, geography, earth science, geology, environment, landscape, management.

Quick crossword 2

Answers on p 300.

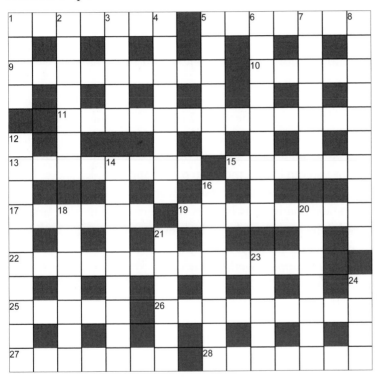

Across
1 Import duties or taxes (7)
5 Ablated (*anag.*) (7)
9 Fruitfulness, richness (9)
10 Semi-aquatic rodent, *Myocastor* (5)
11 Investigates long-term weather (13)
13 Columbia, Athabasca, Dovre, Campo de Hielo Norte (3,5)
15 In line (6)
17 Early film (6)
19 Dispersion, migration (8)
22 Agreed, sanctioned, signed off (6,7)
25 Main blood vessel (5)
26 To make or become acid (9)
27 Weakened, let down (7)
28 stinking chamomile, made yew (*anag.*) (7)

Down
1 Porous limestone (4)
2 Reprocess, salvage (7)
3 Moulds, mushrooms and toadstools (5)
4 Breathing hole, calipers (*anag.*) (8)
5 Decomposes timber (3,3)
6 Earth movement science (9)
7 By the sea, easy bid (*anag.*) (7)
8 or a tequila (*anag.*) (10)
12 Neighbourhood (10)
14 Nomadic traveller (9)
16 Ti 22 (8)
18 Not labour or conservative (7)
20 ... dancing, model it (*anag.*) (3,4)
21 Australia, Madagascar and Man (6)
23 ... council, outhouse (5)
24 Lattice, network, web (4)

15

Researching and writing in teams: it's fun and efficient

How do you know what you think before you talk about it?

Group work is a fundamental part of school and university education. Employers want people who work happily and effectively in teams, because they want to solve problems, issues and challenges that need the co-operation of many people. Teamwork allows the exploration of more material than is possible for an individual, which at university means you can look at more interesting and complex issues. Team members bounce ideas off each other. Where resources are limited, getting involved in co-operative activities and sharing has benefits for everyone. The skill benefits include teamwork, developing professional standards in presentation, group reporting, problem solving, negotiation and responsibility.

A group of students having supper together may talk about Prof. Blownitt's lecture on nuclear power. Everyone has different ideas. If some people decide to chase up further information and pool it, they are beginning

> Where is …? We said meet at 10 after the lecture. Text him.

to operate as a team. In some modules you will research and write as a team, but remember you can use team skills and approaches to tackle other parts of your course. Two to five people in the same flat, hall of residence or tutor group can team up to extend and optimize research activities, and talking about geography or geology together will develop your discussion skills and broaden your views.

This chapter raises some of the issues associated with group work and suggests some ways of tackling group tasks. There are pitfalls. Not everyone likes teamwork and sharing. Some people feel they may be led down blind alleys by their teammates. Overall, I think the advantages of team research outweigh the disadvantages, but if this is an issue that bothers you, brainstorm reasons why certain group members are a disadvantage. Possibly:

☹ those who do not pull their weight;
☹ the perennially absent;
☹ those who do not deliver on time;
☹ the overcritical who put others down, suppressing the flow of ideas;
☹ anyone who gets touchy and sulks if their ideas are ignored.

If you are aiming to benefit from team activities, these are some characteristics to suppress! You may not like working in teams, but when you do, it is in every-body's interests for the team to pull together and derive the benefits. Having identi-fied unhelpful teamwork characteristics, consider what good team qualities might include:

☺ making people laugh;   ☺ making friends;
☺ getting stuff finished;   ☺ keeping calm in arguments;
☺ communicating well;   ☺ speaking your mind;
☺ being reliable;   ☺ resolving disputes.

Reducing tension in a group promotes cohesion and encourages everyone to learn.

There is a natural pattern in life that can also be seen in the normal reaction of people to teamwork and life in general (Figure 15:1). Map these natural reactions to an event in your life, or the response of the Labour Party after its defeat in 2010, or to an assignment, as in Figure 15:1. Each time you are given an assignment, you are likely to experience all these reactions. The people who realize 'where' they are in the sequence, and push on to the recovery and getting-on-track stages, give themselves more time to do the task well.

Natural Reaction	Reactions mapped to an essay	Reactions mapped to a group project
Shock	I know nothing about …	We know nothing about …
Recrimination	I don't want to do this module/essay, why is it compulsory?	I don't want to work with this group/people/at all!
Disagreement	No motivation, not much done.	Everyone is doing their own thing or nothing.
Reorganization	Two weeks to go, time for a plan.	Two weeks to go, we must make a plan for and with everyone.
Recovery	Parts of this are quite interesting.	Everyone knows what to do, some people are enjoying it.
Getting on track	Parts of two sections are organized, and am finding more references.	It's coming together, drafts are being swapped around, and ideas developed.
Partial success	Second draft done, needs editing, good diagrams and references.	Mostly done, but no cover, diagrams missing, references incomplete.
Frustration	Printer needs ink, spell and grammar check to do.	We cannot find Jim who has it all on his USB stick.
Success	Final version handed in. Hooray!	Report completed, time to celebrate!

Figure 15:1 Natural reactions to group 'work'

15.1 Researching in groups

Groups can explore and evaluate more material than an individual, doing more interesting projects, but sharing research ideas and research outcomes can be difficult. People work at different speeds, and have different demands on

> My pre-uni group work was not a good experience. I don't like it. I know where I am if I do something myself, but my module group was great and it was so much fun, getting to know the others ... and I now understand about t-tests because Emily explained them to me. Looking forward to next time.

their time. Group members are only human. Everyone has a different way of researching and note-making; shared information 'looks' different on paper, BUT you can learn from the way other people note and present information. Figure 15:2 summarizes thoughts from three groups after a shared research task. They give an insight into what can happen.

One meeting or three? How often does a group need to meet? Generally: two or more meetings → more chat → greater exchange of views → more interactions → more learning. Have electronic meetings when time is tight. Facebook, VLE chat spaces, Skype and email all help.

Task	Process
1. Tutorial essay on wind power Individual essays Two-week deadline	Met and decided we didn't know where to get data! Split up to do library, online library and www search. Discovered power company brochures and reports. Agreed to keep textbooks and photocopies in Ed's kitchen cupboard where we could all get at them. Middle of week 2 we realized no one had looked at Wind Power alternatives, Sarah did it. Two discussions at lunchtime in the Union, and it became the topic of conversation in Ed's kitchen for a week. Sarah and Mike phoned two companies for further details and got some info, but too late for the company to post things to us before the deadline.
	The research time was about equally split. Generated a bigger pile of articles at Ed's than we could have found alone. Results were a bit general. We needed a better plan at the start or part way through week 1. Found info in the Engineering Library at the end which we should have used, it would have helped. We got lots of electronic source info, which saved time, BUT it cost a lot of printer credits. In future we should be more selective and e-mail to each other to save printing.
	Good to chat, group got to know each other better. One person thought s/he would have started writing sooner, but waited for info from others.

2. Seminar preparation Four people Group presentation for 20 minutes followed by general discussion, class of 25	*We all knew each other and it was easy to share out the reading, we volunteered to cover different bits. We decided to meet 5 days before the seminar, over supper, to discuss who was doing what. Great evening, great discussion, no one wrote anything down, so we had to meet the next day to draft the talk.*
	We had far too much material but we could choose. Took examples from Bosnia (it's topical) and Sudan (map and pictures available). Put other refs on a handout. Decided there were some good www sites and put the site details on the handout.
	Decided to use an 'awkward questions' format. We took it in turns to ask an awkward question, and had two people who tried to answer each one. Changed the questions a bit! Met the day before to practise. Everyone talked for too long, tried to cut bits out. The introduction changed about 25 times!
	The seminar went OK, mostly because we had talked about the points beforehand, so I think we all knew so much more than usual.
3. Organizing a field research day Five people Task to be decided	*Sorting out what and how to do it took ages, there were too many good ideas. We had trouble getting together, and in finalizing the plan. Tended to split into two. The organization of the day was left to Andrea in the end, because she had the car. We met three times beforehand and didn't really decide anything each time. We needed a leader from the start and to focus faster on what to do.*
	The field day was great FUN, Liz and Jamie put the data straight onto the laptop. We changed what we were measuring a few times so by the end we had more records for the sites visited later in the day. Didn't get as much data as we hoped, but talked to some really nice people and the graphs look good. We had loads to write under the 'criticism' and 'what we would do next time' sections.

Figure 15:2 Experiences of group research

15.2 Writing in teams

Team writing is a normal business activity. The same techniques are used in university writing for projects and reports. Start with the group brainstorming an outline for the document and who does what; then research and draft your subsections; circulate drafts for comment; incorporate additional ideas and views; and finally someone edits the final version. Like ensemble recorder playing, writing as a team is difficult and discordant the first time you do it, but a very useful experience.

One valuable approach involves team members volunteering both to research and draft specific subsections and taking agreed 'writing roles' (Figure 15:3). Discussions are more lively and focused when individuals can use their role to offer alternatives and new approaches. The 'critic' has the license to criticize, the 'linker' can say 'yes that's great, but how do we use butter mountains and polar ecology to explain county employment patterns?' Adopting writing roles depersonalizes criticism, which is especially important when working with study buddies. Be critical and stay friends.

Writing role	Responsibilities
Visuals/Graphics	Smart graphs, figures, indexes, contents page.
Summarizer	Introduction, conclusions, abstract.
Critic	Faultfinder, plays devil's advocate.
Academic content	Subject reporters, everyone.
Linker	Checks connections between arguments, sections and the introduction and conclusion.
Discussant	Evaluates and discusses.
References	Checks the selection is complete and in the agreed consistent format.

Figure 15:3 Writing roles

Style and layout

Decide on the general layout of the document at the start, the format, fonts and style of headings, and the format for references. Make decisions about word length, for example 'each subsection has a maximum of 200 words'. Early word-length decisions limit waffle. Nevertheless, length needs looking at later in the project, as some parts of the argument will need more space. Keep talking to each other about ideas and their relative importance and position in the narrative.

Finally, someone (or some two) has to take the whole document and edit it to give it a consistent voice and style, BUT everyone must provide them with the graphs, diagrams and references in the agreed formats.

> It's all feedback. I think that's what I didn't understand when I came to uni. Comments from tutors are feedback, but so are comments and ideas from the rest of the group. Cool.

 TOP TIPS

Group messaging

→ Be polite and friendly.

→ Remember everyone will have different views, opinions and information they are reading.

→ Double check you have attached the right (latest) version of your document.

→ The group writing process is a muddle at times, just as your own research notes are a muddle before you organise an essay structure. Create time to organize and structure the work.

→ Reply to messages within the day, even if is just to explain you will do the next bit tomorrow. Keeping in touch is important for the group dynamic.

Duet writing

Two people sitting at a PC can be very effective in getting words down. The exchange of ideas is immediate, two brains keep the enthusiasm levels high and you can plan further research activities as you go. It is advantageous to have two people together doing the final edit, keeping track of formats, updates and staying cheerful.

Timetabling issues

A team-writing activity cannot be tackled like the traditional essay, in the late evening after a night out. People get ill and things happen, so the timetable needs to be generous to allow for slippage AND team members have to agree to stick to it. It usually works, as it is too embarrassing to be the only non-contributor. Put completion dates against decisions, as in Figure 15:4, and use the project management Post-its, see Figure 4:2, p 43, to create a chart version of this list. Which suits the group better?

1 Brainstorm initial ideas, assign research tasks and data collection (Day 1).

2 Research topics and collect data (Days 2–5).

3 Individuals draft their subsections and share them around the group (Days 6–9).

4 Meet to discuss progress, decide on areas that are complete, assess where additional research is needed, assign further research and writing roles and tasks (Day 10).

5 Individuals redraft sections and circulate them again (Days 10–14).

6 Decide whether we are all happy with this. Pass final version to the editor (Day 15).

> **7** Team members do final checks on figures and references. Check submission requirements are met (Days 16–18).
>
> **8** Finalize cover, contents and abstract. Everyone has a final read. Check and add page numbers. (Day 18).
>
> **9** Submit (Day 18!)

Figure 15:4 Timetabling team writing

Keeping a team on track? Need a chairperson?

One of the pitfalls of working with friends is that more time is taken making sure everyone stays friends than getting on with the task. One vital issue that emerges is the initial division of labour (see Figures 15:2 and 15:3). It is vital that everyone feels happy, involved and equally valued. So the chairman must endeavour to ensure there is fair play, that no one hogs the action excluding others, and equally that no one is left out (even if that is what they want.) The chairman is allowed to hassle you into action. It's unfair to dump the role of chairman on the same person each time, so share it around. Chairing is a skill everyone should acquire.

If a group feels someone is a serious shirker, they may want to invoke the 'football rules', or ask the module tutor to do so. The rules are one yellow card as a warning – two or three yellow cards equal a red card and exclusion from the group. The yellow–red card system might be used to reduce marks. A red-carded person does not necessarily get a zero, but in attempting to complete a group task alone, they are unlikely to do better than a bare pass.

> *We did seven group projects this term. All really good fun when we had sorted out what to do. I had to have a diary for the first time. Lots of meetings. What I learned is that something always goes wrong, NIGHTMARE!, it's just like that. But we got them all done for the deadline.*

TOP TIPS

→ The time required to tidy up, write an abstract, make a smart cover, do an index, acknowledgement page and cross-check the references is five times longer than you think – split the jobs between the team.

→ It saves hours of work if everyone agrees at the start on a common format for references, and everyone takes responsibility for citing the items they quote.

→ The key to team writing is getting the STYLE and TIMETABLE right; CONTENT also matters.

Plagiarism

Where group work leads to a common report then obviously collaborative writing is involved. If you are writing independent reports ensure your reports are independently written, not copies or cut-and-paste versions of each other's documents. In this situation, share reading and discuss ideas, BUT write independently. Plan to finish the research five days before the deadline so everyone has time to draft and revise their own reports.

15.3 Communication by email, SharePoint, Google docs, Skype ...

Distance Learning students do all their group work online, rarely, if ever, meeting face to face. Techniques for sharing documents online are useful for everyone. It is vital that everyone has access to a group report, therefore a method of circulating the most recent version of a document is needed. Using email with attachments is ideal, and saves printing costs. For larger documents there will be a range of new sites by the time this book is published. Currently Dropbox, Google docs and your VLE are all options. Skype and similar platforms are great for discussing the project when you cannot get together. Share thoughts, drafts and updates, while working at the most convenient place and time for you. Have a go at **Try This 15.1** for practice.

TRY THIS 15.1 – Sharing documents by email

Research a tutorial essay or practical report, sharing resources and ideas, without physically meeting. HINT: the first time, it helps to have everyone in the same room so you can resolve any problems quickly. Open a new word-processing document then:

1	Type in the title, the keywords you would use in a library search and write very briefly about two or three issues, say three sentences for each.
2	Save the document. Open email, attach and mail the document to your mates.
3	With a bit of luck, other people will be doing the same thing at the same time, so there should be emails arriving from them. Save your colleagues' files in your own workspace, but check whether they have used the same file name as you, if so it will need changing to avoid overwriting.
4	Return to the word-processing package, open your original document and the new ones. Copy the material into a single file and reorganize it to make a coherent set of comments. While collating, keep track of ideas, perhaps through subheadings. Reflect on your colleagues' feedback. At the immediate supportive level there is the 'that's a new/good/middling idea'. More actively, think around variations on 'I was surprised by ... because ...' 'I disagree with ... because...' 'We all agreed that ... because...'
5	Now look at the style of the responses. What might you need to do as an editor to make these comments hang together?
6	Finally, get together and decide how you will organize your research and writing to maximize the opportunity to share resources and information. You each have a different version so there are issues to resolve.

Bigger documents

Imagine a report constructed by four people, being edited and updated daily. You need a tracking system. Adapting the elements shown below might be a useful template for action.

The document needs:

✓ a header page with title and the outline plan.

✓ a section that everyone updates when they change the document e.g. 'Andy modified Sections 2.6 and 3.2 on Wed 1 April at 10.00'.

✓ an agreed working order, e.g. John drafts Section 2, then Anne revises it. All revisions circulated by day X.

✓ an agreed format for citations and the reference list.

✓ a decision to check email and respond every x days.

✓ use the 'Revision Editor' in your word-processing package. The person responsible for a section will not necessarily welcome five independent redrafts. The 'Revision Editor' highlights revision suggestions for the subsection editor to accept or reject as desired. By opening multiple copies of the document on your PC, you can cut and paste between drafts.

✓ DO NOT USE PAGE NUMBERS TO REFER TO OTHER VERSIONS. These change with almost every version. Use section numbers and date each version carefully.

✓ agree that everyone will keep an archive of all drafts, so that you can revive an earlier version if disaster strikes, people are ill, your computer crashes ...

Use email (Skype, Facebook, MSN ...) to brainstorm ideas among friends, tutorial and seminar groups and old school friends doing degrees at other universities. People like getting messages and usually respond. There are a great many people you could brainstorm with; think more widely than just you and your lecturer.

15.4 Assessment of teamwork

Assessment tends to generate discussion about 'fairness'. There are dark hints about cheerful shirkers getting good marks when their mates have done the work. How is this handled? Most staff will offer some variation on the following approaches for assigning marks, some of which involve team input. If the 'football factor' is in play, there may be an agreed penalty, say −10 per cent for a yellow card.

The simple approach

Each member of the team gets the same mark, so it is up to everyone to play fair.

The private bid

Each individual fills in a form privately, and the assessor tries to resolve discrepancies. This style:

Name ....
Names of other team members ....
I feel my contribution to this project is worth ....% of the team mark because ....
....
....
Signed .... Date...

This approach may lead to discussion amongst your assessors, but it lets people with personal problems acquaint staff privately. It can require the Wisdom of Solomon to resolve. Remember that lecturers always take careful note of what you say, but do not necessarily change the marks.

Team effort

A form in this style asks everyone to comment on the contribution of each team member. Summing the totals constructs an index of activity, which is used to proportion the marks.

Estimate the effort made by each team member 1= no effort, 2 = did a bit, 3 = average, 4 = really useful, 5 = outstanding contribution.	Alex	Dave	Jez	Mick
Attended all meetings				
Contributed ideas				
Did agreed share of the research				
Did agreed share of the writing				
Other information – detail particular contributions:				
TOTAL				

15.5 Comments on teamwork

Lecturer feedback: 'Initially students are very democratic and give equal marks. With experience, students raise the marks of those who have done more of the task'. 'The amount of chat was enormous, and between them they could tackle a more difficult problem than as individuals'. 'The team-writing exercise made everyone think about the order and quality of the material. The report was more extensive and detailed than one person could do in the time'. 'One team was happy to have this year's "mine's a third candidate" because s/he happily did the activities like washing up in the laboratory and making poster backgrounds, while cracking jokes and keeping the team cheerful'.

Figure 15:5 details reflections of first-year students following group research using library and electronic sources. Overall, they see the exercise as positive and the disadvantages surmountable with experience. Their reflections may be worth considering as you start team activities.

Team research and writing usually produces a better document than an individual response. This is because team activities generate a series of drafts and more thinking about the topic and audience, so that the final product is more polished. Make the most of group work activities on your CV. It isn't the academic content that matters; highlight the skills you used in delivering a group product, like negotiation, meeting deadlines, allocating tasks, collaborative writing, editing and co-ordinating research.

15.6 It's all gone wrong!

Stuff happens. University tutors understand this, (NB tutors put a note in your file if you explain your grandmother has died, the record is in double digits). If you have a major commitment during a group project, tell your group at the start. Plan to do your share early on or arrange to do more of the finishing tasks. If there is an emergency, you are ill or have to be somewhere else, make sure you Facebook, e-mail or text the group so that they can help. If your absence will affect the whole group, make sure the tutor knows.

> As a group we worked really well. Sal had to go to her sister's wedding the last weekend before Monday hand-in. She did loads of the research, sorted diagrams and left us with probably more stuff than we would have found. Sweet. The rest of the group did the final editing and checking.

Group work is about sharing tasks and managing problems as they arise. Tutors are deeply unimpressed by people who do not give their group fair warning of problems. In this case they will penalise the individual rather than the group.

Over the course of the degree where you might do 30 or 40 group activities something will go wrong on one or two of them. That is true for everybody. When there is an emergency people are good about reorganizing and coping.

Students completing Assessment Contribution forms are usually ethical and honest when describing the problems they have caused their group, and suggesting where marks should be raised for those people who did the extra work. It is the right thing to do. You will get raised marks when you have helped in a crisis.

What were the advantages of teamwork?	What were the disadvantages of teamwork?
People can help each other to understand. Learn consideration. Lots of prior knowledge got us started quickly.	Recognizing that other people have different ways of working, and having to find a way round it.
New friends/met new people.	Hard to find out if a non-contributor was ill or just being lazy.
Less work per person. Shared ideas.	Who does the writing?
Divided workload saved time.	Difficult to agree on points of view.
Created cohesion amongst five people who had only just met. Greater overview of the subject.	Hard to get a time to meet f2f, OK-ish online.
We covered library, www, CD-ROM and e-journals between us, more than we would have managed alone.	Conflicting views (is this an advantage?). Having to compromise on common topics.
Learning to share and compromise.	
Individuals could concentrate on the bits that interested them. Encouraging to know that others in the team thought you were on the right lines.	

What will you do differently next time you work together?
More thorough planning to concentrate efforts. Set clear objectives at the start.
Improve online sharing to save time. Email each other more often. Read my uni email regularly.
Elect a team co-ordinator, who has the team members' timetables, to find convenient meeting times.
Split the team to tackle different libraries and meet later.
Organize a couple of meetings rather than trying to do the task all in one go.
Reserve books in advance, arrange to meet in library at a quieter time of day. Search electronic sources first, papers and books later.
Co-ordinate reading, so two people do not read the same text.
Start work on the project earlier – not leave doing this to the last three days.
Distribute the workload earlier.

Figure 15:5 Geography students' feedback on a teamwork assignment

15.7 References and resources

Open University 2011 Groups and teamwork, OpenLearn, http://openlearn.open. ac.uk/course/view.php?id=2338 Accessed 15 February 2011

Hay I 1999 Writing Research Reports in Geography and the Environmental Sciences, *Journal of Geography in Higher Education*, Directions, 23, 1, 125–135

LearnHigher 2011 Group Work – Resources for Students, http://www.learnhigher. ac.uk/Students/Group-work.html Accessed 15 February 2011

University of Kent Careers Advisory Service 2011 Team working skills, http://www.kent.ac.uk/careers/sk/teamwork.htm Accessed 15 February 2011

Words in Geo-words 2

How many words can you find in PALAEONTOLOGY?
Answers on p 300.

16

Acknowledging references and other sources

Every piece of academic writing MUST include a reference list, ALL essays, reports, practicals, posters, presentations, field reports ... ALL OF THEM.

There are a number of standard ways to acknowledge research sources; journals use various styles. Some GEES departments have a preferred style, look in your student handbook. LOOK NOW.

You need this chapter if there is limited or no advice in your handbook. Some handbooks give rather minimal advice such as 'References should follow the Harvard System'. This is fine if you know what the Harvard System is. Essentially the advice here is Harvard System with minimal punctuation to simplify your life. This system is used throughout this book, giving you examples throughout.

It may seem pointless but referencing requires consistency in presentation. Decide on a style and stick to it. Consistent punctuation matters.

> **Reference list:** an alphabetical record of ALL the sources cited in your document, whether you read them or not, placed at the end of the text. Do not include 'other things I read, but didn't quote in the text'.
> **Bibliography:** an alphabetical list of sources or references on a particular topic; a complete bibliography would include every document ever written about a topic.
> **Annotated bibliography:** group references into subsections, with a brief paragraph justifying the grouping and summarizing the contents.

Harvard System Lite

1 References in your essay, report ... must give the surname of the author and the year of publication in brackets: Williams (2011) or (Stillwell and Bissell 2010) followed by a, b, etc when there are two or more references to an author for the same year: (Dalrymple 2011c).

2 Page numbers must be given for quotes: (Drake 2010: 42) or Purvis (2010: 178). All direct quotes must be in quotation marks, either single or double ones.

3 At the end of your piece of work all references cited in the document must be listed with authors' names in alphabetical order, and in chronological order where authors have more than one citation.

Your course may introduce you to EndNote (2011) or a similar bibliographic software system that helps you to store information about references and import them to documents. Go to classes to get started with Endnote – it will save time in the future – or use the tutorials on the Endnote website.

16.1 Citing journals and books in your writing

Within your essay or report, cite a book or article by the author's family names and year of publication. When there are two authors, both are quoted, but with three or more authors, the *et al.* convention is adopted. For example, 'Describing the retail geography of Great Buyit, D'Benham (2020) showed that the mega mall developments described by Markit and Hyper (2017) were already unprofitable, whereas the outlet distribution system analysed by McBurgers *et al.* (2015) had already expanded to...'

> PLEASE use *et al.* correctly. Really nice academics are annoyed when *et al.* is misused. It must always be in *italics* because it is Latin meaning 'and other persons', with a full stop after *al.* because al. is an abbreviation of *alia*. See Chapter 28/p 277 xx for further Latin matters.

Where information in one text refers to another, quote both: 'As reported by Stokes (2015, cited in Law 2020) found that ...' Both the Stokes (2015) and Law (2020) references must appear in the reference list. Similarly: 'In an extensive review of kluftkarren, Gryke (2018) shows the field approach taken by Clint (2010) is unreliable, and therefore the methodology adopted by Clint is not followed'. Quote both sources, although you have probably only read Gryke. Quoting both tells the reader how to locate the original. If you want to make clear that you have acquired your information from a secondary source, use a sentence like 'Landslides in North America annually injure 5000 people and cause property damage in excess of $12 billion (Crushum 2015, cited Flattenem *et al.* 2012)'. In this case, it is important to give both dates to indicate the age of the original data, 2005, rather than the 2010 date of the reference you read. You should quote both Crushum and Flattenem *et al.* in your references. The Crushum reference will be cited in the Flattenem *et al.* paper, so not including it is lazy. If Flattenem *et al.* does not cite Crushum, use your library search engines or Google Scholar to find it.

Referring to government publications, where the author is awkward to trace, is a problem. The author may be a Committee, as in many House of Commons publications, or an organization, or the publisher. There are no absolute rules; use common sense or follow past practice.

Referencing by initials can be convenient and time saving. Cite HoC STC (2010), and ensure the initials are explained within the reference:

HoC STC 2010 The Disclosure of Climate Data from the Climatic Research Unit at the University of Anglia, HC 387–I, Eighth Report of Session 2009–10–Volume I: Report Together with Formal Minutes, House of Commons – Science and Technology Committee, UK Stationery Office, London

An Act of Parliament is referred to by its title and date: ' ... in the Apprenticeships, Skills, Children and Learning Act (2009)' and in the reference list as:

Apprenticeships, Skills, Children and Learning Act (2009)–Public General Acts–Elizabeth II, Chapter 22, UK Stationery Office, London

Each year the China Statistics Press publish many statistics books with invaluable data, leading to classic referencing nightmares. The China Statistical Yearbook 2010 may appear as (CSP2010) or (China Statistical Yearbook 2010):

CSP 2010, *China Statistical Yearbook 2010*, China Statistics Press, Beijing;
or
China Statistical Yearbook 2010, China Statistics Press, Beijing.

If there doesn't seem to be a rule, and you cannot find a precedent, invent one and use it consistently.

16.2 Citing paper-based sources in your reference lists

The key is consistency in format: use a standard sequence of commas, stops, spaces and italics. Underline the italicized items in handwritten documents.

Citing a book – the *book title is in italics:* Author(s) Year *Title*, Edition, Publisher, Place of Publication, e.g.:

Pelling M 2010 *Adaptation to Climate Change*, Routledge, London

Citing a chapter in an edited volume – the authors of the chapter or paper in an edited text are cited first, followed by the book editor's details. The *title of the book is italicized*, not the chapter title: Author(s) Year Chapter title, in Editors Names (Eds.) *Volume title*, Publisher, Place of Publication, Page Numbers, e.g.:

Battarbee RW 2010 Aquatic Ecosystem Variability and Climate Change – A Palaeoecological Perspective, in Kernan MR, Battarbee RW and Moss B (Eds.) *Climate Change Impacts on Freshwater Ecosystems,* Wiley and Sons, West Sussex, 15–37

Citing an edited book – Editors names (Eds.) Year *Title*, Edition, Publisher, Place of Publication, e.g.:

Hewitt CN and Jackson AV (Eds.) 2009 *Atmospheric Science for Environmental Scientists*, Wiley-Blackwell, Oxford

There are rare exceptions. For example, there are three editors of The New Rivers and Wildlife Handbook, and a request to cite the RSPB, NRA and RSNC as authors, so it appears as:

RSPB, NRA and RSNC 1994 *The New Rivers* and *Wildlife Handbook*, The Royal Society for the Protection of Birds, Sandy, Bedfordshire

Citing a journal article – *Journal Titles are italicized:* Author Year Article Title, *Journal Title*, volume number, issue number, page numbers, e.g.:

Goodge JW, Fanning CM, Brecke DM, Licht KJ and Palmer EF 2010 Continuation of the Laurentian Grenville Province across the Ross Sea Margin of East Antarctica, *Journal of Geology*, 118, 6, 601–619

In the rare case where there is no author attribution, use the Anon convention:

Anon 2013 Rebuilding Ank-Morpork, *Discworld Advertiser*, 23 May, 4–5

Citing a newspaper article – when an author is cited use: Author Full Date Title, Newspaper, volume number if applicable, page number(s), e.g.:

Jackson J 17 October 2010 How can enlightened societies have institutionalised policies of race profiling? Observer, 22

When there is no author, use the first words of the headline as the cross-reference and put the full headline in the reference list: 'Organic Farm (2012)' in the text. Article Title, Full Date, Newspaper, volume number if applicable, page numbers, e.g.:

Organic Farm Revolutionizes Veg Deliveries, 16 April 2012, The Borchester Echo, 5

Citing an unpublished thesis – thesis citations follow the general guidelines for a book, add 'unpublished', and enough information for another researcher to locate the volume. Note: no italics, e.g.:

Cook VA 2009 Embodied fieldwork: Exploring students' personal geographies of the field, unpublished PhD thesis, University of Leeds

Wetmore B 2015 Late Holocene variations in humidity in Denmark, unpublished BSc dissertation, School of Geography, Earth and Environmental Science, University of Poppleton

Sandy IAM 2015 Coastal Management influences on the beach morphology of the Norfolk Coast, unpublished BSc dissertation, School of Geography, Earth and Environmental Science, University of Poppleton

16.3 Citing e-sources

> NEVER, EVER put a web address in an essay or report.
> The URL ALWAYS goes in the reference list, at the end.

The formats for citing e-sources are settling into a pattern. These notes follow recommendations from various online and library sources. If you are writing for a publication, check the style guide first. The crucial element is adding the date when you accessed the website, because the contents of sites change. Some Wikipedia entries change daily. The next person to access the site may not see the same information. Never put full http://www... addresses within your essay or report text; these only go in reference lists.

Citing web pages – treat internet and other electronic sources like paper-based references. For example '... the estimated and projected HIV/AIDs-related deaths for China (UNAIDS 2012) indicate that ...'

URL addresses tend to be long, they need careful checking. If the citation is longer than one line, the URL should only be split after a forward slash in the address: http://ThecaSe/ofchaRacters/inTheAddress/sHouldnOt/bealterEd.EVER

The pattern is: Author/editor Year Title, Edition, Place of publication, Publisher, URL, Accessed date, e.g.:

UNAIDS 2008 Report on the Global AIDS Epidemic 2008, Joint United Nations Programme on HIV/AIDS, UNAIDS, http://www.unaids.org/en/KnowledgeCentre/HIVData/GlobalReport/2008/ Accessed 15 February 2011

When the electronic publication date is not stated, write 'no date'. The 'Accessed date' shows when you viewed the document. 'Publisher' covers both the traditional idea of a publisher of printed sources, and organizations responsible for maintaining sites on the internet. Many internet sites show the organization maintaining the information, but not the text author. If in doubt, ascribe authorship to the smallest identifiable organizational unit.

Citing online journals – some journals can be accessed directly, some via university library search engines and some via Google Scholar (2011). The route to reading is not important, use the format for hard copy journal articles and add the URL and access date: Author Year Title, *Journal Title*, volume, issue, page numbers, URL, Accessed date:

Lima E and Legey LFL 2010 Water Quality Restoration in Rio de Janeiro: from a piecemeal to a systems approach, *Journal of Environment and Development,* 19, 3, 375–396, http://jed.sagepub.com/content/19/3.toc Accessed 17 February 2011

Citing email – to reference personal email messages, use the 'subject line' of the message as a title and include the full date. Remember to keep copies of the emails you reference.

Sender (Sender's email address), Day Month Year. *Subject of Message*. Email to
    Recipient (Recipient's email address)

McPartlin A (geomcpartlin@poppleton.ac.uk), 21 July 2012. *Essay for Second
    Tutorial*, email to Donnelly D (geoldonnelly@poppleton.ac.uk)

Citing blogs – Author Year posted, Title of blog post, Blog name, Date of posting, URL Accessed date, e.g.:

Romans B 2010 Seafloor Sunday #75: deformation along the Kimmeridge coast,
    clastic detritus, science blogs, 17 October 17 2010, http://www.wired.com/
    wiredscience/clasticdetritus/   Accessed 17 February 2011

Citing wikis – wikis can be very awkward to cite because there are multiple authors, and sites on main platforms such as Wikipedia are updated regularly. For example, the Camelford water pollution incident in 1988 has had updates since Wikipedia began and these continue. The View History link provides the date of the last revision. This particular wiki page has over 500 revisions. By providing your date of viewing another reader can track and view the version that you read, which may have been changed considerably since.
Author Year of last revision, Title of entry, Wiki name, URL, Accessed date

Anon 2010 Camelford water pollution incident, Wikipedia, http://en.wikipedia.
    org/wiki/Camelford_water_pollution_incident   Accessed 12 October 2010

Citing social networking websites – I suggest such references are very rare if you are wanting to be taken seriously. If you need to, use: Author Year Message title, Network site, URL, Accessed date:

Simon C 2015 Answer for fieldtrip group work, Facebook, 12 January 2015,
    http://en-gb.facebook.com/1234567890   Accessed 17 February 2015

Citing Twitter – There are almost certainly better options than a Twitter citation, or use: Author Year Title of Tweet, Platform, Date of Tweet, URL, Accessed date

RSPB Jim 2010 Results for survey, Twitter, 14 September 2010, http://twitter.
    com/RSPBJim/1234   Accessed 17 February 2011

Citing a podcast or vidcast – Author Year Title of Podcast, Platform, URL, Accessed date

If in doubt about the author use a short version of the title in your essay or a sensible shorthand. There is a Harvard geology field trip video on YouTube that makes a good example. You might cite this as (Harvard Geology 2010) in your report, and in the reference list as:

Harvard Geology 2010 Harvard Geology Field Trip to Italy, YouTube, http://www.youtube.com/watch?v=XWszdBLxQ-E   Accessed 17 February 2011

Take time to check the details around any web posting to make sure you are citing as accurately as possible. Authors names and publisher information are often hidden at the foot of pages, or in links embedded in the picture/film/ ...

Citing maps – The first example here is for a map, the second is for a map within an atlas. Citing the scale is important in all map references:

Map maker Year of issue Title of map. *Map series*, Sheet number, scale, Publisher, Place of publication

Ordnance Survey 2009 Blackburn and Burnley, *Landranger Series*, Sheet 103, 1:50000, Southampton, Ordnance Survey

Smith D 1997 Horn of Poverty, Scale not given, in Smith D The State of War and Peace Atlas, Penguin, London, 58–9

See Google Maps API pages for further information on Google map options http://code.google.com/apis/maps/index.html
For Google Maps and Google Earth see http://www.google.com/permissions/geoguidelines.html

By the way, an atlas is cited in the same way as a book:

Dorling D, Newman M and Barford A 2008 *The Atlas of the Real World: Mapping the Way we Live*, Thames and Hudson, London

 TOP TIPS

→ The author's name may be found at the foot of an electronic document. Where the author is unclear, the URL should indicate the name of the institution responsible for the document. However, this organization may only be maintaining the document, not producing it. Take care to assign the right authorship.

→ The date of publication is often at the foot of the page with the author's name, and sometimes with 'last updated' information.

→ Keep accurate records of the material you access, with a database, spread sheet, list ...

Producing correct reference lists is an important skill, demonstrating your attention to detail and professionalism. Can you spot the errors in **Try This 16.1**. The ultimate test of a reference list is that someone else can use it to locate the documents. Check your citation lists meet this standard.

TRY THIS 16.1 – Spot the referencing errors
How many can you spot? Answers are at the end of the chapter.

1 Park C 2001 The Environment: Principles and Applications, Routledge, London

2 <u>World Economic and Social Survey 2010</u> by United Nations

3 Wright RT and Boorse DF 2010 Environmental Science: Toward a Sustainable Future, 11th Edition, Pearson Education,

4 Sumnerd Sedimentology Concepts #4, You Tube, http://www.youtube.com/watch?v=jY3Y7o45Tt0

5 BP 2010 Final Louisiana Bird Release – 3 October 2010, Video, <u>http://bp.concerts.com/gom/lastbirdrelease100310.htm</u> Accessed 17 February 2011

6 Ingram A 2010 Biosecurity and the international response to HIV/AIDS: governmentality, globalisation and security **Area, 293–301**

7 Batchelor, R.A. and Anthony R. Prave 2010 Crystal tuff in the Stoer Group, Torridonian Supergroup, NW Scotland, Scottish Journal of Geology, 46,1–6

8 Fieldwork haihku, Twitter, 14 September 2010. <u>http://twitter.com/gem@.../...</u> Accessed 14 May 2011

9 <u>Natural England</u> (2009) *Sites of Special Scientific Interest(SSSIs)*

10 *Olivella mica,* <u>World Register of Marine Species</u>, Accessed 17 June 2010

16.4 Sources and resources

Anglia Ruskin University 2011 *Harvard System of Referencing Guide*,
http://libweb.anglia.ac.uk/referencing/harvard.htm   Accessed 17 February
2011

Cardiff University 2007 *Citing and Referencing in the Harvard Style*,
http://www.cardiff.ac.uk/insrv/resources/guides/inf057.pdf
Accessed 17 February 2011

Endnote (2011) Endnote, http://www.endnote.com/   Accessed 17 February 2011

Google Scholar (2011) http://scholar.google.co.uk/intl/en/scholar/about.html
Accessed 17 February 2011

Monash University Library 2011 Harvard (author-date) style examples,
http://www.lib.monash.edu.au/tutorials/citing/harvard.html
Accessed 17 February 2011

Staffordshire University 2009 *Harvard Referencing Examples*,
http://www.staffs.ac.uk/uniservices/infoservices/library/find/references/
harvard/index.php   Accessed 17 February 2011

Answers: Try This 16.1 – Spot the referencing errors

1   The name of the book should be in italics: *The Environment: Principles and
    Applications.*
2   This is the most common example of incorrect referencing found in student
    work, the order is wrong and there is missing data. In this example the author
    is also the publisher and the survey that is underlined should be in italic. It
    should read:

   United Nations 2010 *World Economic and Social Survey 2010*, United Nations,
     New York

3   Needs the location for Pearson, and italics for the book title.
4   Is missing date of the podcast, and the accessed date. Should read:
    Sumnerd 2007 Sedimentology Concepts #4, You Tube, http://www.youtube.
    com/watch?v=jY3Y7o45Tt0   Accessed 17 February 2011
5   BP is not defined because BP is a brand in its own right. Originally the
    abbreviation for British Petroleum, and still understood as such, BP became
    the brand name in 2000 signifying 'better people, better products, big picture,
    beyond petroleum'. No italics are needed for a video reference.

6 *Area* is a journal, therefore must be in italics, the page numbers are included but the journal volume number is not. The random bold needs eliminating.

   Ingram A 2010 Biosecurity and the international response to HIV/AIDS: governmentality, globalisation and security, *Area*, 42, 3, 293–301

7 The authors names are incorrect and include extra punctuation, should read: Batchelor RA and Prave AR, and add italics to *Scottish Journal of Geology*.

8 and 10 are missing author and date information.

9 The publisher's details are missing. Brackets have appeared around the date. The author, National England, should not be underlined.

10 The accessed date tells you there is a missing website URL. *Olivella mica* is correctly in italics as a species name. This entry is missing the author and date, and should not be underlined.

17

Effective essay skills

I'm getting on with the essay, I started the page numbers yesterday morning, they look really neat.

Many of the chapters in this book start with some reasons why practising a particular skill might be a moderately good idea, and an analogy involving practising playing musical instruments or sport. Now demonstrate your advanced creative skills by designing your own opening. Create two well-argued sentences using the following words or phrases: reports, communication, language, persuasion, cheerful examiners, lifelong, clarity; with needs . solos, practise, bagpipe, As, writing

The paragraph above may give you an insight into how annoying examiners find disjointed, half-written paragraphs, with odd words and phrases rather than a structured argument. Such paragraphs have no place in essays. If this is the first chapter you have turned to, keep reading!

> *I did History for A level, I understand grammar, my essays are OK, why do tutors keep assuming students can't write?*

> *Last time I wrote an essay I was 12. What is an essay? What do they want?*

What is your experience of writing essays? At university, an essay is usually a 1000-word plus piece of well argued writing. This chapter gives you the important points for essay writing for geography, earth or environmental sciences essay. BUT there are many books and websites that discuss writing skills in more detail. Mature students who have 'not written an essay in years' or those who did science A levels and 'haven't written an essay since ...' should start with this chapter and check various websites, see the selection at the end of the chapter for starting points.

Tutors are often asked 'what a good answer looks like'. The most helpful response involves comparing and evaluating different pieces of writing. This chapter includes some examples of student writing at different standards. You are asked to compare the extracts to get a feel for the type of writing you are expected to produce. Space restricts the selection, but you could continue by comparing essays with friends, study buddies and other members of your tutor group.

All essays need good starts and ends, lots of support material and a balance of personal research and lecture-based evidence. That usually requires an initial plan, some rethinking, writing, further research and rewriting; then a heavy editing session, where the initial long sentences are cut down to shorter ones and paragraphs are broken up, so that each paragraph makes a separate point. The first version of anything you write is a draft, a rough-and-ready first attempt, requiring development and polish before it is a quality product. Most marks disappear because the first draft is submitted as the final product.

Recognize that writing evolves and that GEES matters are never simple. It takes time. There will always be a range of ideas, layers of complexities and examples. This means that as you research and write:
- new data and information will appear and change your ideas;
- early reading matter may need to be excluded;
- new arguments will need to be woven into the text;
- early ideas will evolve and may need to be restated in a new way;
- the examples carry the arguments forward; and
- there is a good balance of space given to the big ideas relative to the smaller details.

17.1 What kinds of essays are there?

The 'tell me what you know about ...' style essay should be disappearing from your life. University questions usually require you to think about information that you have heard in class and researched independently to create a persuasive argument. Questions will ask you to analyse, criticize, examine and debate ideas in a structured way, using apt examples to illustrate your arguments. Interweave lecture material with personal research findings and ideas for high marks. Reproducing the facts and arguments from a lecture may get a mark of 30–50 per cent. Painful but true. For 50 per cent plus you must show the examiner that you have thought about the issues, sorted out what it all means, and put the argument in your own words (Figure 17:1). OK, that's my opinion. Ask your GEES tutors what they think and do.

You will increasingly be asked for discussion rather descriptive essays. Compare these questions:
1 *Descriptive*. Outline and justify the methods of store location you would recommend to a retailer.
2 *Discussion*. Store location is a retailer's minefield. Discuss.
3 *Descriptive*. Describe the role of potassium argon dating in geology.
4 *Discussion*. Evaluate potassium argon dating as a tool for geologists.
The descriptive essay title has pointers to the structure of the answer. The discussion

essay needs more thought and planning; you must establish your own structure, and write an introduction to signpost this for the reader. The introduction leads to a series of linked arguments supported by evidence (*references*), leading to a conclusion that follows from the points you have made. Question 3 lets you explain about one dating method, question 4 requires information on the alternatives to potassium argon dating so that you come to evidenced conclusions about its value. Including material that veers off at a tangent or is irrelevant, or presenting evidence in ways that do not really support your case, will lose marks. Archeologists use potassium argon dating. There are archaeology references and case studies that may be useful or too off the point for this essay.

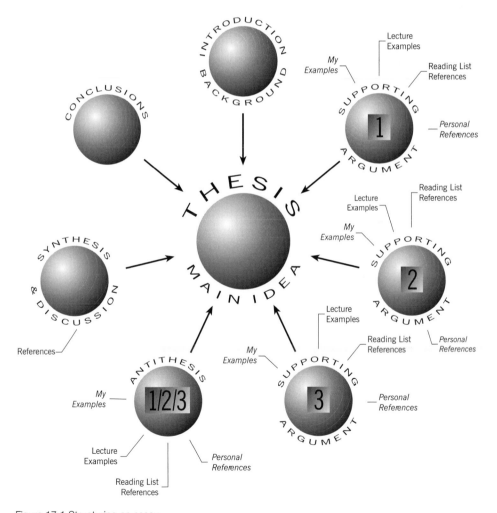

Figure 17:1 Structuring an essay

What are the supporting ideas?

Most GEES essays explore issues with no right answers. You consider the various dilemmas, debates and decisions that are the essence of real situations, provide evidence and reach a balanced conclusion. Questions like 'Analyse the causes of evolutionary mass extinctions', 'Compare the strategies home-

> *My lecturer really means it when he says he wants to be surprised by the answers. He wants 1500 words, references, short paragraphs but the rest is up to me. Scary stuff.*

less men and women develop to survive in urban areas', 'Creationism presents a valuable counterbalance to science. Discuss', 'Consider whether global climate modelling presents an impossible challenge', 'There should be no more nuclear power stations', 'Will focusing environmental and health policies on minority and low-income communities lead to environmental justice?', and 'What is the evidence to support the argument that ecological niches will always heal them-selves?' are all contentious.

17.2 Carefully analyse titles

Take time to analyse the question, it provides guidelines for your answer. For example:

Discuss the causes, effects and methods to reduce hazards associated with earthquakes.

Anyone who thinks that this is an essay about earthquakes is in serious trouble. It is an essay about fire, flooding, landslides, tsunami, collapsing of bridges and disruption to communications – a hazards essay. Earthquakes get a passing refer-ence in exemplification, as for example, 'The collapse of housing stock in the Turkish earthquakes of ...' or 'The severing of network links where roads collapsed during the San Francisco earthquake led to ...'; but they are not the essay's focus.

> *It was sooo stressful, I had no idea what was wanted, trying to fit into this new way of thinking, working. The draft essay we did in first term was v.v.v. useful, not so much mine, but looking at the rest of what my tutorial group did. Between us we shared so much useful feedback.*

Equal weight needs to be given to the three main sections of the essay – 'causes', 'effects' and 'methods to reduce', with each aspect discussed in relation to each hazard.

Describe and critically evaluate the factors that are considered to drive glacial cycles.

What are the key words? The problem with this essay is that there is too much informa-tion. The influence of plate tectonics, orog-eny, the raising of the Tibet plateau and consequent feedbacks to climate, changing global CO_2 and temperature balances, sea temperatures and currents, volcanic dust, astronomical theories, including Milankovitch, equinoctial procession, orbital eccentricity, sunspot and solar flare theories. There is an excess of knowledge material. Most students attempt to include all the knowledge and thereby throw away the 50 per cent of the marks for the 'critically evaluate' section.

Now look at **Try This 17.1.**

>  TOP TIP
>
> ➜ Assume all essay questions include the phrase
> **with reference to specific geographical/environmental/geological examples.**

TRY THIS 17.1 – Key words in essay questions

What are the key words and potential pitfalls with the following titles? Answers on p 300.

1   Discuss the role and responsibilities of science and scientists for society.

2   Communications, internal commerce and energy are the sectors that are usually identified as the most serious 'bottlenecks' in contemporary Chinese development. Explain the weaknesses in one of these sectors.

3   To what extent are growth and change impeded by archaic social structures in either Latin America or a selected country in Latin America?

4   Critically explain how Andrews (1983) derived his bedload entrainment function.

5   Explore the arguments that further drilling for oil should be permitted in Alaska.

6   Evaluate the role of mass media reportage of environmental issues.

17.3 Effective introductions

A good introduction serves two purposes. It outlines the general background position and signposts the structure and arguments that follow. It gives the reader confidence that you are in command of the topic. Short is good, and 100–200 words, about half a side, is usually plenty. If the introduction is longer, consider moving some of the material to later sections.

Introductions require thought. Good opening paragraphs are often written last. Planning time is vital. A provocative headline opening, grasping the imagination of the reader is a good wheeze, BUT it must be integral to the essay. Lengthy case material and examples are not usually good opening material. If there is an example in the first paragraph, limit it to one sentence – further details belong in the body of the essay. Here is a reasonably good introduction (it could be improved):

Explain, with examples, why river flows in urban channels are rarely natural.

Urbanization and the consequent paving of natural surfaces with cement, roofs and tarmac has changed the natural water balance within cities: there is less infiltration, less soil storage and increased rapid surface runoff (Wellis and Mack 2020). Urban flooding can be both personally inconvenient and economically disruptive. Engineering works have channelled runoff efficiently through drains and sewers to minimize surface inundation. The demand for water in the city is high, and in developed countries particularly, "this demand has increased through the twentieth century as lifestyles have changed" (Gluggit 2020: 45). River water is extracted for drinking from local sources where possible reducing flows. Effluent is normally treated within the catchment and returned to the nearest river adding to flows. The competing demands for extraction and disposal of water, combined with maintaining minimum flows in drought periods and controlling flood flows, mean that urban channel flows are managed and regulated rather than natural. These demands are not always compatible; a strategic planning issue for water managers. A range of case studies from the UK and USA illustrate the issues and conflicts that arise.

This introduction summarizes some of the major interlocking issues, indicates that there are no simple answers and that examples will be taken from two developed countries. The Gluggit quote is clearly shown in quotation marks and the author, date of publication and page number is cited correctly. However, the selected quote is not a particularly helpful statement in this example. Use quotes to make strong points.

This next example contains a series of true statements, but the style of the essay is not indicated. AVOID THIS approach please.

Discuss the economic impact of floods in urban areas.

Over history, many very important urban developments have grown up besides or around fluvial channels, which are susceptible to flooding. Noah was one of the first to avoid a flood. Over this same period, the inhabitants of those urban areas by rivers have tried to manipulate the natural channel so as to reduce the risk of flooding and thus economic impact on that area. Although flooding may take place at any point along the channel's path, it is in urban areas that the greatest density of housing is affected, thus maximum disruption takes place and the largest economic impact is felt.

The first sentence is long and woolly; it could read, 'Historically, flood-prone riverside sites have been chosen for settlement'; 10 words, not 20. Similarly, the start of sentence four could, read 'Although flooding may occur anywhere along the channel'; 8 words rather than 12. The second sentence is totally irrelevant and unconnected. There is no indication in this paragraph of the topics the essay will address, no mention of direct and indirect costs, insurance and liability, planning legislation to control floodplain development, modelling insights from hydrological and economic aspects and case examples.

A more headline-style start might read:

> *Who pays if your business is flooded by storms, snowmelt, hurricanes, water backed up from overloaded storm sewers, or the failure of a dam or embankment? More than 90 per cent of the disastrous floods in the US are not declared as federal emergencies. Flooded properties do not usually qualify for financial assistance. Many UK residents do not have flood insurance. If your business is flooded, the costs are your problem ...*

or

> *You do not have to live on a floodplain to be flooded in an urban area. Flooding following unusually large storms, the backing up of sewers, rapid snowmelt, collapse of floodwalls, dams or embankments, or the failure of a mains water supply pipeline can hit any home or business...*

Pretend to be an assessor and take a critical look at **Try This 17.2**.

TRY THIS 17.2 – Evaluate an introduction

Pretend to be an examiner looking at these four introductory paragraphs to the same essay. You need no knowledge of Canada or Wetlands; a gap in your content knowledge helps you concentrate on the main issues. Questions to ask include:

- Is the general case outlined and explained?
- Does the introduction indicate how the author will tackle the essay?
- Is the language suitably technical (or 'grown-up', as a tutee once described it)?

The essay title is horribly general, which makes a good opening particularly important. See comments on p 302.

Discuss the impact of man on Canadian wetland landscapes in the last 100 years.

Version 1

Wetlands are described as transition zones between terrestrial landforms and water bodies. They resemble both uplands, due to their ability to support emergent plants, and aquatic regions, because of the domination of the areas by water. Even though these highly productive areas contribute to 14 per cent of Canada's total land surface, they were not considered initially to have much value because they made transport difficult across Canada, whilst at the same time harbouring insects that led to many deadly diseases.

Version 2
Wetlands perform a range of valuable functions, including reducing flooding, filtering polluted waters, defending shorelines and as habitats for many species, including migrant birds, which makes them landscapes worth retaining. They are found wherever the water table is at or near the surface for most the year and include salt and freshwater marshes, bogs, sloughs, fens and the margins of lakes and rivers. They are dependent for their structure and function on a sustained hydrological and nutrient status, but this also makes them vulnerable. A change in the water supply through drainage, drought or flooding will upset the ecology, as will a sudden increase or decrease in nutrient supply (Verboggi 2020). Recognition of the value of wetlands is a recent phenomenon. In the past they were considered as wastelands, dangerous quagmires that could trap animals and the habitat for insects that cause diseases such as malaria. Early settlers in Canada could afford to ignore wetlands – there were more promising lands to farm – but as the pressure on land grew through the twentieth century, wetlands were increasingly reclaimed. In the past 100 years a diversity of competing interests has led to the reclamation for agriculture, drainage schemes, urban expansion, forestry, and peat extraction for horticulture and fuel (Albertus and Brit-Columb 2020). The competing issues and impacts are illustrated here through a range of case examples, from Arctic to continental wetlands, highlighting the vulnerability of the sites and the problems of protection and reclamation. However, despite the loss of 70 per cent of Canada's wetlands, there are good examples of conservation and site protection, exemplified through RAMSAR (2011) and other sites.

Version 3
Due to the ever-increasing population and rapidity of land use change, the prevalence of wetland areas (wetlands are a transition between landforms and water bodies) is depleting. Within urban areas wetlands are often the last to be developed, and as cities have expanded through this century there has been a rush to fill these in too. It is man's impact that, particularly through the last century, has influenced them most. Many have been depleted due to their site near to main rivers, coasts and bays, and therefore have been developed due to their necessary use for transportation. The conversion of lakes and reservoirs has added to this. This essay will examine some of the impacts on wetlands in the last 100 years.

Version 4
'Canadian wetlands are extensive but vulnerable to exploitation and reclamation' (Environment Canada 2020:123). The pressures of agriculture, drainage schemes, urban development, forestry and peat harvesting have significantly reduced their extent and this has had an impact on the wildlife which use wetlands as permanent or migration habitats. This essay will assess the extent and processes of change, using examples from wetlands across Canada, and then focus on the particular issues raised by the Copetown Bog near Ontario. As this case reveals, there are grounds for optimism as the significance of wetlands (RAMSAR 2011), and the need to conserve them, is nationally and internationally recognized.

17.4 The middle bit!

This is where to put the knowledge and commentary, developed in a logical order, like a story. You need a plan. Creating a list of bullet points, putting them in order and writing a paragraph for each is a great plan. Subheadings make plans clearer to you and the reader. Whether you use subheadings mentally, or mentally and physically, is a personal matter. Ask your tutor if s/he minds subheadings in essays.

Examples are the vital evidence that support the argument. Where you have a general question, 'Describe the impact of satellite TV on global marketing', aim to include a range of case examples, BUT if the question is specific, 'Comment on the impact of Sainsbury's marketing strategy', then the focus is primarily on Sainsbury's. In the 'global marketing of TV' essay, remember to make the examples global. There is a temptation to answer questions like this with evidence from the USA, Europe and Japan, forgetting the developing world. Use as many relevant examples as possible, with appropriate citations. Generally, many examples described briefly get more marks than one example retold in great depth. Examples mentioned in lectures are used by 80 per cent of your colleagues. Exhilarate your examiner with a new example!

Graphs, figures, maps and pictures

Graphs, figures, maps and pictures illustrate and support the arguments in fewer words than a description of the same material. Put them in. Label them fully and accurately. Refer to them in the text. If there is no instruction to refer to Figure 1, the reader gets to the end without looking at it, and the inclusion is wasted. Clearly labelled figures can be cross-referenced from anywhere, especially the discussion and conclusion sections.

Reference appropriately

All sources and quotations must be fully referenced (see Chapter 16 for details), including the sources of diagrams and data. Avoid the over-obvious: 'Deforestation affects carbon stocks and prices' (Strassburg 2009). OK, Strassburg may have said so in a lecture, but this is not really an appropriate example of referencing. Whereas 'Results show that at low CO_2 prices (US\$ 8/t CO_2) a successful mechanism could reduce more than 90% of global deforestation at an annual cost of US\$30 billion' (Strassburg *et al.* 2009: 265), is a proper and highly desirable example of referencing.

17.5 Synthesis and conclusion

A conclusion is the place you have reached when you are tired of thinking.

Synthesis

The discussion or synthesis section allows you to demonstrate your skill in drawing together the threads of the essay. You might express the main points in single sentences with supporting references, but this is just one suggestion. Here are two examples of synthesis paragraphs. In answer to the essay question: 'Evaluate the principle theories of infiltration', paragraphs 2–8 outlined the main theories with the equations, the knowledge section of the answer. Followed by:

> *Current understanding of the processes of infiltration is influenced by the three main theories outlined above. For dry and sandy soils with high permeability, the Green and Ampt (1921) model is appropriate. The Phillips equations (1937) have been widely and successfully used, for example Engman and Rogowski (1976), in an overland flow routing model. In saturated and semi-flooded soils, Hillel's (1986) model provides a more accurate set of results than the Phillips approach. This is because permanent or semi-permanent ponding on the surface reduces the air infiltration element from the equation. Neither of these models is appropriate in permafrost areas and with frozen soils. For these environments, the results of Dunne (1964), although limited to soils in Nebraska, seem to provide a more accurate model. In addition to these field approaches, the data from laboratory experiments by Jackson (1967) and Burt (1982) have improved our process understanding in detail. However, these approaches, although apparently successful, have yet to be tested at the field scale. At present there are a suite of approaches available to the hydrologist. However, s/he needs to select the technique that is most compatible with prevailing environmental conditions.*

So what do you think? Lots of information, interwoven references, useful information – this should get high marks and is a style to aim for. The last three sentences are very good. If the facts were right it would get high marks. BUT this particular paragraph will be lucky to score 40/100. The authors are all hydrologists and some did the work referred to, some did not and all the years are wrong. What is worse, the most recent reference is 1986 – has nothing happened since then? This is an unacceptably dated answer for 2011+. Examiners notice spoof and imaginary references.

Conclusions

The conclusion should sum up the arguments. Do not repeat statements from the introduction or introduce new material. This causes marker's to comment: '*The conclusions had little to do with the text*'. It is hard to judge the standard of conclusions in isolation, because the nature and style of a conclusion depends on the content of the essay, but it helps to look at other examples. I would advocate reading articles in *The Economist, Nature, New Scientist, Observer,* and *Time*. There is something relevant in most issues. Despite the caveat above, criticize the examples in **Try This 17.3**.

TRY THIS 17.3 – Good concluding paragraphs

Read the following concluding paragraphs and consider their relative merits. Feedback on p 302.

Version 1

It is therefore possible to see that man has had a mega impact on the landscape as the land devoted to agriculture quadrupled and there was an explosion in urban development. This impact was caused by the discovery that wetlands have many important values and perform many worthwhile functions. The components of the system are seen to be interconnected, and destruction of one part of the system will have affects, probably adverse, elsewhere in the system. This has meant that for the last 25 years there has been less damage, but only a great deal of time and conservation will reverse the extraordinary exploitation that has occurred on Canada's wetland landscapes.

Version 2

All in all, there has been a lot of pressure on Canadian wetlands over the past 100 years due to man's impact, but things look set to improve, with talk and actions of restoring many wetlands areas, and also new international regulations and laws for their protection.

Version 3

In addition to the impacts I have considered, threats likely to continue or increase include the harvesting of black spruce (*Picea mariana*), use of wetlands for wastewater treatment, burning, which may lead to the loss of native species and the invasion of Eurasian weed species, and increased accumulation of sulphur and heavy metals from acid rain. The impact of man has led to the loss, or serious degradation, of these vulnerable ecosystems, which are effectively a non-renewable resource. The recreation and educational value of wetlands has been discussed. There is a dilemma for managers here, who need to protect the ecology of sites while giving appropriate access for eco-tourists, hikers, ornithologists, fishermen and hunters. Good *et al.* (1978) pointed out that much wetland management derived from common sense rather than science. In 1999 the science base has increased, the complexities of ecosystems interactions are better understood. The RAMSAR Convention (2011) has provided valuable guidelines for wetland preservation, but there is still much to be done to prevent further wetland destruction.

Version 4

Canadian wetlands were drastically altered by the intervention of humans. Seventy per cent have been reclaimed for a variety of purposes, including agriculture, urban development, industry, energy development and harvesting. There is a consequent loss of habitat and reduced species diversity. Recent changes in human use of wetlands have led to some abandoned agricultural sites being returned to wetland, but agriculture is still the dominant activity.

Version 5

To conclude, I think it important to highlight the damage done by man to wetlands, both on a national scale and at a local scale, where only small amounts of interference can cause potentially devastating effects. This essay has shown that without greater management of these environmentally sensitive areas, man will not only destroy a vast educational resource but also affect the entire global atmospheric system.

17.6 Assessment

Assessment

Either use your department's essay feedback form or Figure 17:2 to critically self-assess your first and second drafts. The results will indicate where to focus your next research and writing efforts. In Figure 17:2 the percentage equivalents are guides which vary between assignments.

Essay Title						
	1	2.1	2.2	3	Fail	
Information 30%: topic covered in depth; multiple examples; embedded references						Superficial content; few examples; few references
Argument 40%: logical presentation of ideas; one main idea per paragraph; good use of illustrations – diagrams, graphs ...						Disorganized ramble; few illustrations
Creativity 20%: includes some new element – idea, presentation of data, graph ...						No evidence of personal thinking about the ideas
Presentation 10%: correct grammar and spelling throughout; clear layout; quotations in "..."; complete and accurate reference list						No or few references; spelling and grammar errors

Figure 17:2 Self-assess an essay

Another way of developing your evaluation skills is to use **Try This 17.4**, which builds on an idea from note-making by using codes to locate the different sections of an essay and to compare their relative weights. **Try This 17.5** presents another way of creating essay structures, by posing questions about the mark distribution. If the balance of the reward changes, then so does the content.

⚙ TRY THIS 17.4 – Analysing essay structures

Analyse an essay you have written recently using five coloured pens, or a code system. Mark in the margin the sections which show Knowledge, Analysis, Synthesis, Evaluation, and Creative abilities. Now think about the balance of these five parts. Where could the balance of the essay be altered to improve it?

- How would you rewrite this essay to increase the analysis section?
- How would you redesign to increase the evaluation content?
- Look at the structure of your essay:
  - is there a good opening paragraph?
  - does the argument flow logically?
  - are the examples relevant?
  - are the arguments summarized effectively?
  - are the conclusions justified?

⚙ TRY THIS 17.5 – Marks for what?

Take any essay title (the last or next) and work out a plan for researching and answering the question when the mark distribution changes:

1 Knowledge 20%; Analysis 20%; Synthesis 20%; Evaluation 20%; Professionalism 20%; AND

2 Knowledge 10%; Analysis 50%; Synthesis 20%; Evaluation 10%; Creativity 10%; AND

3 Knowledge 30%; Evaluation 70%.

Getting advice

Showing a draft essay, or any document, to someone else for comment is not cheating. It is normal practice in business and academic publishing, as shown by the acknowledgements at the end of many published papers where there are thanks for reviewers' feedback. You do not have to take note of the comments, but having an independent check on grammar, spelling and someone asking awkward questions about content does no harm. Do it early so that you can revise your essay.

> Always number pages

17.7 Getting the English better

Try not to cross anything out! Keep sentences short wherever possible. Practise with **Try This 17.6**.

⚙ TRY THIS 17.6 – Shorten these sentences

Answers on p303.

Wordy	Improved
In many cases, the tourists were overcharged.	
Microbes are an important factor in soil processes.	
It is rarely the case that sampling is too detailed.	
The headman was the proud possessor of much of the land in the vicinity of the village.	
Moving to another phase of the project ...	
Chi-square is a type of statistical test.	
One of the best ways of tackling prison reform is ...	
The investigation of cross-bedding at Ramsgate continues along the lines outlined.	
The nature of the problem ...	
Temperature is increasingly important in influencing the rate of snowmelt.	
One prominent feature of the landscape was the narrow valleys.	
It is sort of understood that ...	
It is difficult enough to learn about stratigraphy without time constraints adding to the pressure.	
The body of evidence is in favour of ...	

Technical terms

Good professional writing in every discipline uses technical language, defining technical terms when necessary. Definitions can be very important: housing policies in Leicester are rather different from those in Sao Paulo; an intense rainstorm in Singapore is considerably more intense than one in Glasgow; a socialist, liberal or conservative government means different things under different regimes.

There is no need to define technical terms that are in everyday use and where you are using words in their normal sense; assume the reader is intelligent and well educated. In an essay on hydrological flows in river basins, paragraph 2 started with: 'Water covers ⅓ of the earth's surface, mostly as oceans. Oceans are expanses of saline water which vary in dimensions, depth and width, and form e.g.: calm, rough, warm, cold'. In my opinion, this student wasted time on the second sentence – an examiner knows what an ocean is. Keep the language suitably technical and the sentences simple. Use the correct technical terms wherever possible and avoid being unnecessarily long-winded, as in 'snowmelt is largely a seasonal phenomenon, which occurs when temperatures rise after frozen precipitation has fallen'.

Startling imagery

'It is a truth, universally acknowledged, that a man in possession of a large screen satellite TV on Cup Final Day, must be in want of a six pack'. The impact is greater for being the antithesis of Austen. However, this comment, while arresting, is dangerously stereotypical, far from politically correct and possibly gendered. Creative use of language is great; avoid the temptation to bowdlerize or tabloidize inappropriately.

One idea per paragraph

In a lengthy essay, restricting paragraphs to one idea plus its supporting argument should make your message clearer. Use **Try This 17.7** to analyse one of your own essays.

TRY THIS 17.7 – One idea per paragraph

Reread your last essay underlining the ideas and supporting statements. If each paragraph has a separate idea and evidence, award yourself a chocolate bar and cheer. If not, redesign a couple of paragraphs to disentangle the arguments and evidence. The challenge is to make the evidence clear to the reader by separating out the different strands of the argument.

NB This is a good exercise to do at the end of the first draft of every essay.

Synonyms

Part of the richness of English comes from the many synonyms that add variety, depth and readability to writing. Take care you write what you mean to write! A common error is to confuse *infer* and *imply* – they are not synonyms; *infer* is used when drawing a conclusion from data or other information, *imply* means to suggest or indicate. We might infer from a dataset that the world is flat, or a lecturer might imply during a discussion that the world is flat. Have a practice with **Try This 17.8** or **Try This 17.9**. Look at any sentence you have written, and play around with the thesaurus settings or book to find synonyms you might use. If you tend to overuse certain words, make a list of synonyms and substitute some of them, WHERE RELEVANT.

Which of the two synonyms makes sense in their geographical, environmental or geological context? Answers on p 303.

1 Studies on litter and organic matter dynamics also aid awareness/understanding of nutrient mineralization.

2 My research project uses focus group interviews as its main/paramount data source.

3 The discrimination/distinction and mapping of rock types are a major focus of remote sensing in geology.

4 River networks can be deduced/extracted from satellite images, allowing controls on drainage patterns to be analysed.

5 What economic logic elucidates/explains plant closings and firms' exit from industries?

6 Modelling is particularly important if real appraisals/estimates of the extent and timing of hazard events are to be of practical value.

7 Past experience/practice of negotiating emission agreements suggests that, despite legislation, success in emission control is by no means assured.

8 Any process by which a marine-ice sheet may form must overcome the problem of enhanced calving rates associated/comparable with deeper water.

9 There is some limited documentation of the effects of channel and water management on the morphology/shape of channels in the Mediterranean region.

Argument by analogy

Analogies can be very useful, but this example shows they can get out of hand: 'A fast breeder reactor is more efficient than a normal nuclear reactor. It can be thought of as a pig, eating anything, rather than a selective racehorse'. The first sentence would be better standing alone. 'An unconformity is where two beds of rocks are out of order, like a sponge cake with the jam missing'. I do not think the jam element helps. Be more technical.

I know what I meant!

Tutors struggle to encourage students to reread and correct written work. Rereading reveals illogical statements and where crucial words have been missed out. Here, for your amusement, are some essay statements that do not exactly convey the writer's original intention: 'For many centuries cars have been causing problems on the roads' (when were they invented?). 'No data can be gathered as the area is relatively flat'. 'Trace fossils are thought to have existed as early as very early on'. 'Large-scale windmills should be located in areas where winds blow almost inconsistently'.

TRY THIS 17.9 – Synonyms again

These tables show a perfectly acceptable sentence and some of the synonyms that might be substituted. Some synonyms are acceptable, others make no sense at all. Which are useful substitutes?

All	research	involving	people	as	participants	has	an	ethical	dimension
	study	including	family		assistants			decent	scope
Any	inquiry	comprising	humanity	because	contributors			equitable	range
Each	examination	containing	tribes	during	helpers	owns		worthy	compass
	encompassing	groups		colleagues				size	
Every	investigation	entailing	nations	while	associates		a	just	sweep
Whole	scrutiny	embracing	races		collaborators			upright	limit
			society						

Landscape	and	time	are	the	subject's	key	concepts
Countryside		duration			theses	primary	ideas
Vistas		past			argument's	main	notions
Perspective		future			contents	foremost	thoughts
Panorama		span			topic's	index	inspirations
Scenery		cycle			point's	catalogue	images
Environment		season			substances	first	impressions

Abbreviations, acronyms and txt spk

Using well-known and established acronyms (see p 287) and phrases like NIMBY (Not In My Back Yard) is fine, but replacing long words or phrases with initials or abbreviations is regarded as lazy by some tutors. When using abbreviations, the full definition must be given the first time the phrase is used, with the abbreviation immediately afterwards in brackets. Biogeographers and botanists conventionally adopt an abbreviation strategy when citing the Latin names of plants (see p 280).

No text speak please. Der r no occasions whr wrds hwevr specl cud tel yr story, or wil 2 gt mrks frm a tutor. Njoy txtng @ oder ty/ms W Fs.

Using nouns as verbs

Please avoid transforming nouns to verbs as in 'to podium' or 'to novelize', or verbs to nouns. Avoid 'the rurality of the site' (use the rural site) or 'the malnutrite stock' (the stock suffering from malnutrition) and similar phrases. Making up new words is not good.

Colloquial usage

Please avoid regional or colloquial terms as they may not be understood. Writing as you speak is also a trap, as in these examples from student's essays: 'He therefore put to greater emphasis on the results of ...' It should read, 'He therefore put too great an emphasis on ...' And 'The large events will have always permeated through the pervious ground', rather than 'the large events will always penetrate through ...'

Google 'commonly misused English words' for more examples.

> Avoid *cool, wicked, maybe, neat, so dead,* and *bingo* as descriptors of good things.

Spelling and punctuation

Spelling is potentially a minefield: use the spell checker, and then check carefully. The following sentences have errors that were not picked up by a spell checker : 'the Haddock shopping centre' (the example was Haydock), 'models can be said to be lumped or disintegrated', 'Fare Trade', 'focussed on the shear pace of change', from the census we calculate the morality of inhabitants','middlemen can charge exuberant prices', 'the competition between predators and pray'.

The excessive use of exclamation marks!!!!!!!!!! is not good! And please never start a sentence with And, But or BUT. There are plenty of examples of its (strictly) incorrect usage in this book, where BUT and capitalization are used to emphasize points.

Check that you do not overuse certain words. Find synonyms or restructure the paragraph where repetition is a problem.

17.8 References and further reading

CCC Foundation 2011 Guide to Grammar and Writing, http://grammar.ccc.commnet.edu/grammar/index.htm   Accessed 15 February 2011

This is a very jolly, visually rich site. Watch the Americanisms (unless you are in the US), but worth a visit. See especially:

Plague Words and Phrases, http://grammar.ccc.commnet.edu/grammar/plague.htm Accessed 15 February 2011

Crystal D 2004 *Making Sense of Grammar*, Pearson Longman, London

Greetham, B 2008 *How to Write Better Essays*, 2nd Edn., Palgrave Macmillan, Basingstoke

Guildford, C 2011 Paradigm online writing assistant, http://www.powa.org/ Accessed 15 February 2011. (A clear and jolly site.)

Holtan, R 2011 How to build an argument, and how to write an essay, http://homepages.ed.ac.uk/rholton/write/writehome.html   Accessed 15 February 2011

Jackson, D 2010 How to write a critical essay, www.kent.ac.uk/english/writingwebsite/writing/article4_p2.htm   Accessed 15 February 2011

Open University Learning Space 2010 Essay and report writing skills, http://openlearn. open.ac.uk/mod/oucontent/view.php?id=398972&direct=1   Accessed 15 February 2011

Russell, S 2001 *Grammar, Structure and Style*, 3rd Edn., Oxford University Press, Oxford

Strassburg B, Turner RK, Fisher B, Schaeffer B and Lovett A 2009 Reducing emissions from deforestation – The 'combined incentives' mechanism and empirical simulations, *Global Environmental Change*, 19 2, 265–278

University of Victoria 2011 The Essay, The UVic Writer's Guide, http://web.uvic. ca/wguide/Pages/EssaysToc.html   Accessed 15 February 2011. An incredibly comprehensive site

Geojumble 2

Find the three GEES words in these jumbled letters. Answers on p 304.

18

Practical reports, laboratory and field notebooks

You are only as good as the group behind you

This chapter is for all GEES students including the human geoggers. Skills involved in report writing include clear thinking, analysis, synthesis, written communication and persuasive writing. The test of a good report is:

> I submitted a virtual report last week.

> Oh, OK, thanks, I'll give it a virtual mark.

✓ The reader should be able to repeat the work without reference to additional sources.
✓ The reader should understand the significance of the outcomes of the work in its wider context.
✓ It is short and to the point.

Some practicals are truly investigative, where data are analysed to find an 'answer'. You might use census data to forecast migration patterns for the next 20 years, or water chemistry analyses to explore pollution issues. The emphasis is on achieving and evaluating the answer. This type of practical is possible because you did 'developmental' practicals. *Developmental* practicals help you understand the research process, by evaluating where and how each step works. In this kind of investigation the report places less emphasis on the final result than on the steps involved, making statements about assumptions, validity of data, techniques, sources of error and bias, alternative approaches used and discarded, or evaluated and rejected. There may be discussion of the choice of statistical test, a forecasting model will be taken apart and its assumptions and processes questioned, alternative field approaches such as questionnaires versus interviews discussed. This kind of practical calls on your analytical thinking as well as your problem-solving skills. Depending on the type of practical investigation, your report will change its focus and emphasis. Many dissertations include elements of developmental practicals, questioning the processes and procedures used.

> Support arguments with evidence

18.1 The format

The normal expectation is that the writing is brief and direct, and follows the general format outlined in Figure 18:1.

Chapters/Sections	Contents
Title	Short and precise title.
Authors details	Author's name, student numbers and contact address (email).
Abstract	Short, precise summary.
Introduction	Background. Why you are doing this.
Methodology	Computer, field and laboratory procedures, sampling methods, site details.
Results	A summary of the findings using tables, graphs and figures.
Discussion and evaluation	Interpret the results using statistics and modelling as needed. Consider the accuracy and representativeness of the results and their interpretation.
	Discuss their significance in both this experiment and their wider geographical relevance.
Conclusion	A brief summary of the outcomes. Describe what could be done to continue the research.
References	Vital – remember to include methodological references too.

Figure 18:1 Outline for a report

Abstract or summary

You might write this in the style of an executive summary (see p 201). Explain the broader context (where this work sits within palaeogeomorphology/youth studies/spatial analysis), the hypotheses tested, methodology, main findings and interpretation in brief sentences.

Introduction

Describe why you carried out this piece of work and give a brief indication of where it fits with the rest of GEES knowledge. The hypothesis being tested should be stated precisely. Refer to other experiments and research on this topic, which may involve a mini (1–5 paragraphs) literature review.

Methodology

The technical details describing the research approach, computer programs, field or laboratory procedures.

Site details: A short site description and reasons for choice of site will set the scene. Use maps, grid references, sketches and photographs so the reader can locate and picture the location.

Sampling methods: Outline the sampling and analytical scheme. Include maps, flow charts and equipment diagrams. This is the point to discuss potential errors, calibration issues, and the influence that the selected sampling procedure might have on the results.

> What is the hypothesis? State it clearly.

Standard and non-standard procedures: Many computer programs, statistical and laboratory procedures are standard. They must be cited, but do not need to be explained in detail. Statements like 'Questionnaire responses were coded and entered for analysis using SPSS', or 'The standard procedure for water hardness by titration was followed (Geography Laboratory Manual Test No 65)', or 'The program was written in Grass and imported to ARCINFO for testing', will do fine. You do not need to tell the readers about SPSS, how you tested the water, or give details of GRASS or ARCINFO. BUT if you make up a new method, adapt a standard method or adopt a non-standard approach, it must be explained in enough detail for the next researcher to evaluate and copy your technique. A full technical or laboratory protocol might be put in an appendix. Explain why it was necessary to invent a new method; what was the problem with previous approaches?

Please note: Science students have a habit of correctly including laboratory procedures, chemical lists, apparatus details and sample size information, BUT omitting details of the subsequent qualitative, graphical or statistical analysis, which loses marks!

> Be clear about definitions:
> **Accuracy:** how close a measured value is to the actual (true) value.
> **Precision:** how close measured values are to each other.
> **Bias:** systematic (built in) error which makes all measurements wrong by a certain amount.

Results

Summarize the results in tables, graphs, flow diagrams and maps to show the main findings. Label and order the diagrams so the sequence matches the logic of the discussion and evaluation sections. Consider putting the raw data in an appendix, or submit on disk; don't clutter this section with it.

Discussion and evaluation

This section will cross-reference to the results section, with the emphasis on explaining relationships, lack of relationships and patterns in the results. Consider how the results fit with previous researchers' experience. Refer back to the introductory and literature sections to place these results in their wider sub-discipline context.

Caution: There is usually more than one explanation for most things. All measurements are prone to error and inaccuracy; if you have not worked out what might be a problem with the data, talk to people to develop some ideas. No self-respecting examiner will ever believe that you have a perfectly accurate data set. Unless you are absolutely certain, it might be wise to use a cautious phrase like, 'It would therefore appear, on the basis of this limited data, that ...', or, 'Initially the conclusion that ... appears to be justified, but further investigations are required to add evidence to this preliminary statement'.

Conclusion

Resist at all costs the temptation to explain all the Wonders of the Known World from two measurements of pH in a mixture of mashed leaf, a $1\,cm^3$ rock sample, or a simulation of the 'Changes in population of the village of Pigsmightflybury 2012–15'. Ensure your conclusions are justified by the data, not just what you hoped to find. This is the place to suggest what research should be done to continue the investigations.

References

A report is a piece of academic work. Reference your sources and quotations fully (see Chapter 16).

Appendices

Some information is too detailed or tangential for the main sections, and distracts the reader from the exciting storyline, but it may be included in appendices. Appendices are the home for questionnaires, details of laboratory protocols, example calculations, programs and additional data, maps and graphs. Appendices are not 'dumping grounds' for all raw data and field notes. Limit appendices as far as possible. For most laboratory reports they are totally unnecessary.

18.2 Laboratory notebooks

Many GEES practicals are assessed through laboratory reports or notebooks which are written during the practical and submitted as the laboratory class ends. This approach encourages the adoption of professional laboratory practice, noting experimental problems as they happen and making interpretations at the time. Getting good marks probably, almost certainly, indubitably, means reading handouts and doing the recommended pre-practical reading, so that you can interpret the results as you obtain them. A practical class is short and this approach seems tough, but it develops your skill in describing and analysing results at speed, describing what you actually see, rather than a fuzzy vision partially recalled three weeks later. If you've never completed a laboratory notebook here are some basic guidelines.

Picking a notebook

Select a large (A4) hard backed notebook. A mix of lined and blank pages is useful. Some modules may specify notebooks with carbon sheets or graph paper. Number the pages. It's important practice to cross out an entry you are unhappy with; do not tear out a page. Lab notebooks are written in pen so that it is clear what are your first and later records.

Content

Page 1	Name, email address, student number, module name and number or project name.
Pages 2–3	Table of Contents, add to it at each session.
Pages 4–5	Leave blank at the start. It gives you space to add a summary, introduction or key as needed.

After that aim to record experiments, starting each one on a new page and cross-reference to the contents page.

Recording experiments

Primarily laboratory note-keeping involves being organized and careful. Ensure you have enough detail so that another student or your tutor can follow exactly what you did, knows the dates and times you worked, and that there is a record of problems and issues. For some modules your work will be supervised, and your laboratory record signed as a correct record of activities by your tutor, laboratory technicians or fellow students.

If you are not given other advice, follow a consistent format for each entry: title; introduction and aims; methods and procedures; data obtained; issues and problems; summary of outcomes.

It is good practice to record more detail than you will need for your report. You are keeping records so that you can be very precise about instrumentation, time of day, other people helping ... so that if there is a problem you can refer back to your notebook to check exactly what you did.

Before you leave the lab take five minutes to read through your laboratory notes, check that you have consistent records, that each experiment has been introduced and the outcomes summarized, and that the basic information is all included. It is very easy to forget to put dates on entries, to include the main results but not the equipment calibration results – everybody forgets things.

Laboratory notebooks are used because they are excellent practice; they help you develop as a scientist or social scientist.

18.3 Field notebooks – geologists especially

All GEES students make notes in the field. Students on geology courses will get special training in their use because professional geologists use their field notebooks as their professional record of activities. These are very important documents which *must* be accurate; they are the evidence if a client disputes a geologist's field results. All GEES students benefit from understanding the value of keeping a professionally accurate fieldwork notebook. Getting practice as an undergraduate is very useful for professional work in later life. Creating good notebooks gives you practice in summarizing information. This is particularly relevant if you want a career in earth sciences, forestry, planning, garden and landscape design, ecology, and the construction industry, and useful in any activity where keeping records is important.

On some geography and environmental sciences field classes your field notebook will be assessed; on geology field trips it will almost always be assessed. Remember it's the notes made in the field that will be assessed. Modules in Earth Sciences in particular give you the opportunity to improve your field sketching and summarizing skills. Expect a tutor to ask to look at your field notebook during a field class, to show you how to improve your sketching style and note making. Really good field teachers will complete their own field notebooks during a field class to show you how a professional geologist, environmental scientist or geographer makes records at a field site. Take advantage of any opportunity to see a professional's field notebook. (You will discover that most people are really bad at sketching, but are making an enormous number of notes on each diagram.)

Your main ambition is to make records, draw sketches, make notes of features, people and activities, and to slot in photographs so that you or somebody else visiting the site at a later date understands what you have done, and can build on your results.

Picking a notebook

Select a notebook that is hardbacked for writing notes in any conditions. Make sure it fits in your jacket pocket, and has a waterproof cover because it will rain at some time. Use pencils, which will write on wet days and are good for sketching, as well as pens and erasers. Ink runs in the rain, pencils are vital. Find a large clear plastic bag, big enough so you can get your hand inside to write in your notebook when it is really wet.

Contents

Some modules will have very specific rules about the way in which the field note-book should be set out. In the absence of handouts telling you exactly what to do, number all the pages and aim for:

Page 1　　　　your name, contact email and address (a surprising number of field notebooks get lost), module number, field location and dates.

Pages 2–3　　Leave blank initially to add contents list later.

Pages 4–5　　Leave blank initially to add material later. You usually want to add some sort of introduction or summary.

For each visit/location you need: location information including map or GPS reference; date; names of people for group work and weather conditions as stand-ard items.

You may want to write a short paragraph outlining your aims for the session/visit, the hypothesis to be tested ... Then you need as much information as possible so that you can interpret it at another time.

Field sketches get better with practice. The trick is to make sure that every picture is very well labelled. Practise landscape sketches, cross-sections and 3D sections as appropriate to your site. Before you leave a site check that you have the map or GPS reference for your location, a compass direction arrow, an idea of the scale, slope angle if appropriate, and details of each feature which is important. If you use abbreviations on your sketches make sure there is a key to explain the abbreviations in either the front or back of your field notebook.

TRY THIS 18.1 – Start drawing

Ford (2003) has a very good exercise to help you get started in drawing.

Follow the web reference Ford C 2003 Sketch Practice 1 and 2, School of GeoSciences, University of Edinburgh, http://www.geos.ed.ac.uk/undergraduate/field/fnb/sp1.html and http://www.geos.ed.ac.uk/undergraduate/field/fnb/sp2.html　Accessed 20 February 2011.

Have a go.

In complicated landscape it is better to have five sketches, each showing a particular feature with its components identified, than one very confused sketch.

As with all skills, keeping good field notebooks develops with practice. People's ability to create sections and cross-sections really improves after a few visits to quarries or cliffs where the geological structure is obvious. Field sketching can be practised every time you see an urban or rural landscape.

18.4 Where the marks go!

Most students would like marks to be awarded for all the time spent sweating over a steamy PC, talking to billions of shoppers and persuading them to complete questionnaires, trying to log into remote computers, digging trenches across moraines and mixing sewage sludge in acid. TRAGICALLY, most academics see these efforts as worthy of minor rewards. A tutor produced handouts, briefings and explained what to do. Practical sessions develop your skills and experience in handling data, samples or spades, and are worthy of maybe 50 per cent of the marks at the most generous. What your tutor really wants to know is:

- Have you understood what the results mean?
- Do you know why a geographer or geophysicist would want to be doing this in the first place?
- Do you understand the flaws in the method and the accuracy of the results?
- What would the intelligent environmental scientist research next?

> Know the deadlines!

Most marks go for the **discussion**, **evaluation** and **conclusions** section of the report. Leave time to write these up and redraft them so the writing is good and *add the references*. Many reports lose marks because they stop at the results stage, missing the interpretation.

> Share drafts to get feedback from friends and study buddies.

This next sentence completed a student report: 'It can therefore be concluded that a fuller investigation is required, increasing both the sample size and its spatial extent'. This conclusion was on the right lines and showed the author appreciated that sampling was an issue BUT it is possible for the author to add value with a conclusion that is more information rich. Something like: 'The restricted data set has limited the inferences that may be drawn. Follow-up research should incorporate a broader cross-section of respondents. If finance were available, extending the research to contrast small towns (Bridgwater, Colchester, Ledbury and Thirsk) with major industrial and commercial cities (Bristol, Glasgow, Leeds and Manchester) would increase confidence in the conclusions'. This is longer but no more difficult to write. The original author of the first statement explained she had discussed future experimental work informally, earlier in the project, but said 'I was shy of adding these sorts of ideas to the report'. Shy can be nice but doesn't get marks. Forget 'shy' when reporting.

Use your departmental practical assessment sheets as a guide to self-assess and revise your reports, or use Figure 18:2. A 60/30/10-mark split is indicated here; what counts in your department?

	1	2:1	2:2	3	Fail	
Structure and Argument (60%)						
Logical Presentation						Discontinuous report
Topics covered in depth						Superficial report
Links to (broader discipline) clearly made						No links made to (broader discipline)
Clear succinct writing						Rambling, repetitious report
Technical Content (30%)						
Experiment appropriately described						Experiment poorly described
Graphs, tables and maps fully and correctly labelled						Incorrect/no labelling of graphs, tables or maps
Sources fully acknowledged						No references
Correct, consistent use of units						Inconsistent use of units
Presentation (10%)						
Good graphics						Poor use of graphics
High standard of English						Limited fluency in writing
Feedback						

Names ....
Laboratory Report Title ....

Figure 18:2 Self-assess laboratory report

Drawing conclusions

It can be very tempting to draw firm conclusions from your data. It's important at the end of a project to think about how much evidence you have accumulated and its wider value. Any GEES student project looks at a limited picture because there is neither the time nor the resources to work on the issue for a number of years. Check the phrases used in your discussion and conclusion. If you're using words from the first column, might the conclusion be better expressed using phrases from the second column?

Unarguably we can see	This limited evidence suggests
Certainly	Possibly/dubious
Unquestionably	Perhaps
Absolutely confirmed	Lend some evidence to
Positively conclude	Tentatively conclude
Definitely indicates	Lends cautious support to

18.4 Further reading and resources

Fox C 2009 The logbook, or laboratory notebook, http://courses.essex.ac.uk/CC/CC301/logbooks.html   Accessed 15 February 2011

Hay I 1999 Writing Research Reports in Geography and the Environmental Sciences, *Journal of Geography in Higher Education*, Directions, 23, 1, 125–35

Kitchin R and Tate NJ 2000 *Conducting Research into Human Geography: Theory, methodology and practice*, Prentice Hall, Harlow

Imperial College London 2010 Working In The Laboratory/Your Laboratory Notebook, http://www3.imperial.ac.uk/physicsuglabs/secondyearlab/workinginlab Accessed 15 February 2011

Drop Out 2

Remove one word from each column to make one word, and the remaining letters align to make six additional words across. Answer on p 304.

M	A	X	M	O	T	H
M	E	M	M	I	A	N
P	U	R	I	R	E	E
G	E	M	T	E	A	U
P	L	A	T	O	P	N
I	S	O	T	E	A	E
G	E	O	D	C	S	Y

19

Reviewing a book

Authors love criticism, just so long as it is unadulterated praise.

Reviewing a book allows a broad theme to be considered. With luck, a book review assignment involves a text with only five zillion pages and a two-week deadline, which adds to the fun (possibly). Admit it, given six weeks you would put it off until the last two weeks! Read some book reviews in journals in the library or online. Try *Green Muze* (2011), *Economic Geography Research Group* (2011), *Environmental Economics* (2011), or *The Scottish Journal of Geology* (2010).

Book reviews are highly personal, reflecting the opinions of the reviewer. A good review provides an objective summary of the book's contents, scope, treatment and importance. You are aiming to evaluate its quality, limitations, applicability and to make comparisons with alternative texts. This means reading competitor books and papers so that your evaluation is fair. Helpfully GEES textbooks have index and chapter headings to help focus your thinking.

Start reading well in advance. Your brain needs time to develop opinions. Use the SQ3R technique (see p 60), especially if time is short. Make notes as you read. These could include:

- a brief summary of the contents;
- the author's aims for the book and the intended audience;
- quotations or references to new ideas to illustrate the review;
- a brief summary of the author's qualifications and reference to his/her other texts;
- the texts that this book will complement or replace;
- thoughts on any significant areas omitted.

All books have their strong points. If you find it weak in places or imbalanced then say so, but counterbalance this with praise for the strong sections. Aim for an impartial review. Use quotes appropriately and reference them at the end. Including long quotations is not a good idea unless they really illustrate a point. Read other texts; ask yourself how they compare. Including examples and evidence from alternative texts will add richness and balance.

Remember to think constructively about what is included and more widely about what is omitted. Authors have to make choices and usually justify them, but it helps the reader to know both what is and is not included.

People read reviews to find out which books to read, they like to know whether it is well written. Where undergraduates are the audience, will they find it easy to read? If it is well written please say so: 'this text is clearly written, with case studies and examples illustrating key points', or 'although intended as an undergraduate text, its style is turgid; the average undergraduate will find the detail difficult to absorb'.

Is the book good value? You can check the price on the publisher's or Amazon's website and compare its price to alternative texts. If it's a good read then: 'Great read, buy it now'.

TOP TIPS

→ Begin by skip-reading the book!

→ Make notes about principal themes and conclusions; then re-read to check these are right.

→ Think, check your notes; decide on a theme for the review.

→ Draft the outline to support your chosen theme; read the book to check that nothing vital is missing.

→ Draft the review; think and check your notes.

→ Edit and revise the final version, add references.

The real skill in reviewing involves giving yourself enough time to absorb the content of the book and letting your brain make connections to other pieces of reading so that your links, compliments and criticisms are compelling.

> A book review **can't** be completed in one draft on the night before a tutorial!

19.1 References

Economic Geography Research Group 2010. Book Reviews, http://www.egrg.org.uk/bkreview.html#bk2010   Accessed 15 February 2011

Environmental Economics 2011 Book Review blogs, http://www.env-econ.net/book_reviews/   Accessed 15 February 2011

Green Muze 2011 Green Book & Film Reviews, http://www.greenmuze.com/reviews.html   Accessed 15 February 2011

Scottish Journal of Geology 2010 Book Reviews, http://sjg.lyellcollection.org/
content/current   Accessed 15 February 2011

Add up 3

Work out the totals for the top of the pyramids. Answers on p 304.

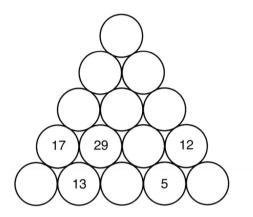 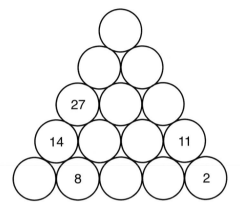

Abstracts and executive summaries

I got the abstract sorted yesterday ... so that just leaves the dissertation.

Abstracts and executive summaries (ES) inform readers about the contents of longer documents. Writing an abstract or an ES enhances your skills in reading, identifying key points and issues, structuring points in a logical sequence and writing concisely.

Most GEES journal articles start with an abstract that summarizes the contents. Executive summaries are normally found at the start of reports and plans, particularly with business documents. An ES aims to describe the essential points within a document, usually in one or two pages. Depending on the context, the style may be more dynamic and less formal than an abstract.

20.1 Abstracts

Look at some well-written abstracts before writing one. Most journal articles start with the abstract, see *Energy and Environment, Transactions of the Institute of British Geographers, Hydrology and Earth System Sciences*, or *Journal of Geophysical Research* for examples.

An abstract should be a short, accurate, objective summary; there is no room for interpretation or criticism. Abstracts should do the following:

☺ let the reader select documents for a particular research problem;

☺ substitute, in a limited way, for the original document when accessing the original is impossible;

☺ access through translation research papers in other languages.

Now do **Try This 20.1.**

If you are asked to prepare an abstract of some text as a tutorial exercise, aim to:

✓ give the citation in full;

✓ lay out the principal arguments following the order in the text;

✓ emphasize the important points, highlight new information, omit well-known material;

✓ be as brief, but as complete, as possible;

✓ avoid repetition and ambiguity, use short sentences and technical terms;

✓ include the author's principal interpretations and conclusions;

✗ do not add your own commentary. This is not a 'critical' essay.

Aim for about 80–150 words. The first draft will probably be too long and need editing.

TRY THIS 20.1 – Abstracts

Next time you read a journal article, read the paper first and make notes without looking at the abstract. Then compare your notes with the abstract. Are there significant differences between them? Think about how you can use an abstract as a summary.

Remember: reading abstracts is not a substitute for reading the whole article.

20.2 Executive summaries

An effective executive summary is a shortened version of a document aimed at a more general audience. A business may produce an ES for the general public, local authority or promotional purpose. ESs are generally less literary than abstracts with bullet points or numbered sections. The general rule on length is one side only, on the basis that really, really, really busy people will not read more. For promotional purposes an upbeat, clear style, with lots of impact is advantageous.

An ES for a student report or project is normally one side or about 350 words. Eliminate all extraneous material. Do not include (severely limit) examples, analogous material, witticisms, pictures, diagrams, figures, appendices, and avoid repetition. An ES should:

✓ be brief and direct;

✓ include all the main issues;

✓ indicate impacts, pros and cons;

✓ place stress on results and conclusions;

✓ include recommendations with relevant costs and timescales.

Look at the discussion and conclusion sections for the main points. Find examples of ES through **Try This 20.2**. Geography students, set the task of generating an ES thought it looked remarkably like a good outline plan, albeit with the emphasis on results and recommendations.

TRY THIS 20.2 – Executive summaries

Search the web for ES examples using 'executive summary', adding any topic you choose. Adding 2011 will prioritize examples from that year, or start with:

Deloitte 2010 Deloitte Volunteer IMPACT Survey, 2–4, www.deloitte.com/view/en_US/us/ About/Community-Involvement/volunteerism/50eed830bee48210VgnVCM200000bb42 f00aRCRD.htm   Accessed 17 February 2011
(This ES shows how a company views workplace volunteering.)

World Economic Forum 2010 Executive Summary; Improve the State of the World: Rethink, Redesign, Rebuild, Davos-Klosters, Switzerland 27–31 January 2010, www.weforum.org/pdf/AM_2010/AM10ExSummary.pdf   Accessed 17 February 2011

Industry Taskforce on Peak Oil & Energy Security 2010 Second report of the UK Industry Taskforce on Peak Oil & Energy Security (ITPOES), 6–7. http://peakoiltaskforce.net/wp-content/uploads/2010/02/final-report-uk-itpoes_report_the-oil-crunch_feb20101.pdf Accessed 17 February 2011

United Nations Environment Programme, 2009 Support for Environmental management of the Iraqi marshlands 2004–2009, Executive summary 7-10, www.unwater.org/ worldwaterday/downloads/Support_for_EnvMng_of_IraqiMarshlands_2004-9.pdf Accessed 17 February 2011

On p 72, the use of abbreviations was suggested as a way of speeding up writing, dealing with longer words, or with phrases repeated regularly. How do you view the use of ES rather than executive summary in this chapter? Should this style be adopted? Note that when the phrase first appears, (ES) occurs afterwards to indicate that this abbreviation will be used thereafter.

Words in Geo-words 3

Set 8 minutes on the timer (cooker, mobile ...) and see how many words you can make from:

FIELDWORK

Answers on p 304.

Dissertations and extended projects

The First Law of Dissertations: Anifink tht kin goo rongg, wyll.

Most UK degree courses have a final year module that involves personal research and an extended piece of writing. Whether it is called a dissertation, project, long project or something else, it will count for 20–60 credits and therefore can have a considerable impact on your final degree result. Each department has its own timing, style, preparation, expectations of length and monitoring procedures. Most departments produce detailed guidelines and run briefing sessions so everyone is aware of the rules. Missing these briefings is a BAD idea; one of your better ideas will be to reread the module handbook every six weeks, to remind yourself of expectations, milestones and guidelines.

This chapter does not in any way replace or pre-empt departmental guidelines and advice. It does aim to answer some of the questions asked by students in the first two years of a degree when a dissertation is sometimes viewed as a kind of academic Everest, to be assaulted without aid of crampons or oxygen, and to say something about timescales for effective planning. More detailed information will be found in Flowerdew and Martin (2005) and Knight and Parsons (2004). If you are going overseas, start with Nash (2000a,b) and Scheyvens and Storey (2003). Skills addressed during dissertation research include autonomous working, setting and meeting personal targets in research, professional report production and problem solving. Compile your own skills list to add to your CV as you undertake your research.

21.1 A dissertation is not an essay, so what is the difference?

Essentially, a dissertation is an opportunity to enquire systematically into a topic or problem that interests you, and to report the findings for the benefit of the next person to explore that material. Some dissertations are published, see, *Reinvention* (2011), *Geoverse* (2011) and the *Plymouth Student Scientist* (2011). Aiming to produce a product that is worthy of publication is appropriate. It is an opportunity to explore material, develop an idea, conduct experiments, analyse information and draw mature conclusions from the results. The results need not be

> *It was amazing, if v scary to do just what I wanted, and to go to Namibia too.*

mind blowing, discovering a new continent or geological era is rare, and the world will not end if your research has a completely expected or completely unexpected answer.

Giving yourself time is vital. Most dissertations start in second year to give ample thinking and action time.

21.2 What is the timescale?

Typically, you will explore a topic by yourself, making all the decisions about when, where, how and in what detail to work. Supervisors will advise, give clues as to what the department expects, but basically it is down to you to plan and organize. Look at the handing-in date, say 1 May of final year, now work backwards.

1. Allow three weeks for slippage, flu, visitors, Easter, career interviews, despondency (April).
2. Writing-up time – allow 3–4 weeks – OK, it should take 6–7 days, but you have other things to do in March (March).
3. Analysing the data, fighting computer systems, getting print-outs, creating graphs .... (February).
4. Recovering from New Year, start-of-year exams, career interviews (January).
5. End-of-term balls, parties, preparing for winter break, end-of-term exhaustion, Christmas vacation (December).
6. Get data from the local authority, environment agency, census, field collection, laboratory analysis ... allow 6–8 weeks (October and November).
7. Have an idea, check out the library, think about it, talk to supervisor, have two more ideas ... settle on a topic, field visit, pilot survey, allow 6–18 weeks (start up to a year earlier).

A final-year project or dissertation, designed to be done between October and May, will consume chunks of time through the period. With an early October start date (meaning late October, because weeks 1 and 2 are needed for recovering from the vacation, starting new modules, catching up on friends and partying) a 1 May finish is very close. Start planning in Year 2. Can fieldwork or data retrieval be

> I had this great idea, got the field site and discovered they had built a housing estate all over my river. Right Muppet moment.

> I got to Tunisia, then realised I had no clinometers.

completed in advance? Can you check out potential field sites in Easter vac of Year 2? Summer vacation fieldwork, and especially fieldwork abroad, needs advanced planning, the right equipment, clear formulation of hypotheses, and a workable, planned timetable.

A dissertation should not be rushed if it is to get a high mark. Most low marks for dissertations are won by those who start very late, and those who leave the crucial thinking elements to the last week. You may have noticed useful thoughts occurring a couple of days after a discussion or meeting. You wake up thinking, 'Why didn't I say ...' High marks become attached to third rather than first-draft dissertations. This means being organized so that there is thinking time at the end.

21.3 How are topics chosen?

Aim to explore an idea or hypothesis that fits your subject interests *and* captures your imagination. If you are not interested, you are unlikely to be motivated to give it lots of time and thought. If your future life appears to involve panning for small change from cardboard city, consider a dissertation that is relevant to possible areas of employment, demonstrating your interest and skills to a future employer and giving you an insight into petroleum product marketing or pollution control. A dissertation may present an opportunity to indulge a hobby, mountain biking, or to visit someone, BUT must address a good question, as in 'the impact of mountain biking on soil erosion', or 'immigrant labour in Toronto suburbs, (where my aunt lives)'.

Questions that start, How ... ? To what extent ... ? Which factors cause ... ? help to focus thinking. Coming to judgements can be more difficult, as in 'Are Keynsian economic values more useful than ... ?' Avoid topics that are very big and largely unanswerable like, 'Will geography have relevance in economic planning in the second half of the twenty-first century?' Steer clear of topics where data are impossible to obtain, for example 'A discussion of the geology and landscapes of Pangea at latitudes 25–45°N', or, 'An evaluation of the geomorphological history of the sixteenth moon of Saturn through analysis of the geochemistry of surface samples'.

 TOP TIP

→ Think small at the start.

Spotting gaps

Throughout the degree course, make notes of thoughts like: 'Why is it like that?' 'Is it really like that in the supermarket, town, glacier, desert, landslide, quarry that I know?' 'Is that really right?' 'But I thought...' Also, when lecturers say, 'but this area is not researched', or 'this was investigated by x in 1936, no developments since'. Another entry point is when an argument seems to have gone from alpha to gamma without benefit of beta. Now, it may be that the lecturer does not have time to explore beta on the way, and there may be a great deal known. Alternatively, beta may be unexplored, a little black hole in need of your torch.

Having spotted an apparent 'hole', take a couple of hours in the library and online to see what is available. Many a tutor has sent a student to research a possible topic, to be greeted by the response that 'the topic is not on because there are no references available'. SUCCESS IS FINDING LITTLE OR NOTHING. This is what you want, a topic that is relatively unexplored so that you can say something about it. Researching for an essay you want lots of definitive documentary evidence to support arguments. For a dissertation you want to find little, or contradictory material, to support the contention that this is a topic worth exploring. If there is lots of literature, that's OK, use it as a framework, a point to leap off from, to explore and extend. Watch too for areas of the subject where the published literature becomes out of date very quickly. Anything to do with government policy, such as housing, changes with a new party in power, and can change rapidly within the lifetime of an administration as policy evolves. You can investigate the current position.

Browsing is a primary dissertation research technique. Immerse yourself in material that is both directly related and tangential to the topic. Wider reading adds to your perspectives and should give an insight into alternative approaches and techniques. You cannot use them all, but you can make the examiner aware that you know they exist.

Topics

There will always be many questions to be answered. One or two topics are done to death. There are fads that relate to whatever is taught in the two weeks before dissertation briefings, and standard chestnuts often expressed as, 'I want to do something on air quality; faulting; global warming; water chemistry changing downstream; dolphins; crime in the city; social housing'. These are quite positive statements compared to the student who wants to do 'something historical' or 'something to do with weather' or 'something while canoeing in Scotland'. The choice of dissertation is *your* responsibility. There may be a departmental list of suggested topics, but thinking of topics and deciding on a valid research hypothesis is your business.

Following a personal interest might lead to an extremely good dissertation on 'badger habitats', 'the environmental impact of the Felphersham bypass' or 'the impacts of EC legislation on the fishing industry in the 2000s'. You will get lower marks if you bias your answer with anti-badger-baiting literature only, your diary from 'my summer as an eco-activist', or present an entirely pro-Greenpeace fishing story. Writing any of these stories would be fine in a newspaper article that seeks to put across a single viewpoint. A dissertation is an objective academic exercise and therefore requires objective reporting. If you feel extremely strongly about an issue, you might write an excellent or a direly unbalanced report. Think about it.

Survey-based studies

Most GEES projects involve collecting and evaluating data. Hence most degree courses include modules on document analysis and survey techniques, including quantitative analysis, questionnaire design, observation and interview techniques (Kitchen and Tate 2000). Results are reported through quantitative and qualitative summary of the responses and statistical analysis where appropriate. Where data are required on a longer historical scale, secondary data, such as the census, national or government data, national and international statistics and library archives, are vital sources.

Collecting your own data has the advantage that you know where and when it was compiled, you have a feel for its accuracy and know exactly why you asked particular questions. The disadvantages include the time-consuming nature of repetitive sampling and some consideration of sample representativeness is required. The time available for data collection is short; data are often representative of one day, week or month in one season, at a limited number of sites. This is better than no data, and a perfectly good way to proceed. Be wary of over-generalizing from the individual to the global case. Avoid inferences like 'The 30 grains of sand monitored in the first order tributary of the River Swale moved 6mm in 12 weeks last summer, therefore we can conclude that River Nile sediments have moved an average 16 miles since the birth of Christ'.

The critical issue with secondary data is access. Data always takes longer to arrive than you hope. Can you discover why it was collected originally? The purpose of the original collector, and your purpose, are almost certainly different. This is not a problem, but you need a paragraph somewhere to tell the reader you understand how the original data collection methods and collation processes impact on the results. Think about data decay issues too: does your data characterize the period you are interested in, or is it x years old? It is tempting to match the best secondary data available with current observations and not to address the data age gap. It may not matter or it might be really important, depending on your topic. Ask whether the arbitrary nature of sampling influences results. For practical reasons, sampling is undertaken on an hourly, weekly, bi-monthly, etc. timescale; does this introduce bias in the results?

Case studies

Exploring 'Social diversity in two villages in Eriador' or 'Managing flooding in Ankh-Morpork' or the 'Fault patterns on Gallifrey' can be great. Case studies allow detailed examination of a specific area, time-period, incident, event, cross-section or computer simulation. The aim is to understand more about how the world normally operates by examining a typical example. Alternatively, you might want to explore the exceptional case to see how the general model responds in extreme circumstances. Watch out for conditions where your 'typical study' deviates from the average.

Time series or cross-sectional studies

Time series studies let you examine the way different groups respond to the same phenomena or to a changing situation. You could look at shopping patterns in people in each 10-year age cohort. This allows you to explore, and compare behaviour and attitudes, to differentiate patterns of shopping in the under-10s, teenagers and older cohorts. Take care to remember the influence on different cohorts of cultural and social influences. You may wish to explore changing attitudes and conditions with individuals, thereby generating longitudinal information.

Downstream, down slope, down fault ... studies of sediments, water quality, soils ... and weather pattern change over time are regular topics. Ensure the background control conditions are kept as constant as possible. It should be no surprise to discover bedload chemistry changes downstream if the river passes a number of geochemically distinct bedrocks, two mines and three sewage treatment works on the way.

Theory-based topics

An examination of the geographical, environmental or geology theory related to a part of the subject is perfectly acceptable and potentially very rewarding, though may seem a little risky at first glance. What will not get high marks is a voyage into your personal opinions on 'the Weberian approach to social class', or a personal diatribe on the ecological impacts of mine waste tailings. Tackling a theory-based dissertation allows you to think through a series of related issues, but there must be evidence to support the argument. This requires careful inspection of information. Be especially careful to locate the supporting and detracting arguments, and to come to a balanced judgement about the full range of material. Set the theory in its discipline context; this requires wide reading around alternative positions.

Before embarking on a theoretical study, talk the ideas through with a tutor. S/he will be enthusiastic and have good advice about the ways in which you might develop your thinking. Having read and thought further, go back and discuss the new ideas. Theoretical work needs to be bounced off people, so talk to your house-mates, tutor group and other GEES-minded people, including postgraduates if possible. Share your ideas and become familiar with them. Browsing and discussion are essential elements of the theoretical geographer's research methodology.

Developing or evaluating a technique

This makes a good dissertation if careful attention is paid to sample size and getting the statistics right. You might look at the accuracy of an instrument; at variations on mathematical equations; at a sensitivity analysis of part of a simulation model; or at the comparative accuracy of measuring something with three or more techniques. The advantage is that the problem is defined fairly tightly and, provided enough measurements are made, the precision of the results can be evaluated.

What if ... project

This is a speculative dissertation, and requires sensible thought and valid forecasting frameworks. It must be a topic where you can gain some information, whether from data or historical expertise, to build the argument, and the counter-argument. 'An evaluation of the geographical consequences of re-nationalizing the UK railway system in 2015' would be a speculative piece of research, embedded in an examination of the historical consequences of the last nationalization of the railways and commenting on issues surrounding privatization. Reporting a number of 'what if' scenario forecasts using transport and economic models will ground the speculations and arguments. This type of dissertation demonstrates synthesis skills.

Beware

There is a temptation to feel you should show one thing or another, that the issues are black and white. Most environmental matters are complex with subjective and objective disagreements. Much of what we know is an approximation; researchers hope to make the best inferences, given the information-gathering procedure available at the time. Interview-based information is subjective, relying on people telling the truth and there being time to collect enough material. Even objective data has its inaccuracies. Population geographers can supply you with data on a country's population, but they will also tell you that even where the census is well regulated, individuals avoid the count for personal reasons, and errors occur in collection and calculation procedures. The estimated historical population of England in 1777 is just that, for 977 and 477 'population facts' are debatable estimates. When exactly did the Holocene start, never mind the Cretaceous period? Knowledge is partial; go for your best interpretation on what you know now, but keep looking for and recognize areas of doubt in your arguments.

A tutor is important, especially in warning against falling for trite arguments that can appear beguiling. Watch for moral arguments which are acceptable in one culture or time, but inappropriate in another culture or time; evaluating third-world development polices using the standards and attitudes of a twenty-first century northern European will not work. Arguing that something 'is the best' is also fraught with danger. There is a good case to argue that Humboldt is the greatest geographer, but that accolade, I argue, personally, goes to Captain Cook. Some less well-read persons might want to give this credit to Johnston, Haggett or ... (these two sentences could keep a tutorial group arguing for some time). These last few sentences are extreme, unbalanced and unsupported as they stand. Who is the TOP environmentalist or earth scientist? There are many candidates to discuss. Avoid unsupported arguments and strong language unless there is over-whelming evidence.

Ethical issues

All research involving people raises questions around the ethics and rights of the participants and the researcher. Departments and universities have their own guide-lines. If you are working with people, interviewing participants directly or on the telephone, using video or observing group interactions, follow the guidelines; they may affect the way you structure your research programme. There are ethical issues about confidentiality, obtaining consent to use the material, photographs, publish-ing, anonymity for respondents and action to take if participants decide later they want to withdraw their comments. You may also need to protect yourself from accusations about non-ethical behaviour by having an impartial observer with you while you are speaking to participants and recording the conservation.

21.4 What about resources?

Is the data you need available at the scale you want, and in the timeframe available to you? There may be plenty of Spanish customs documents for trade in the nine-teenth century, but if they are locked in a Cádiz office and you do not read Spanish, they are no good to you. Finding out about the availability, scale and accuracy of the data that you would like to use is a task to complete early. If the data does not exist at the right scale and in enough detail, your project will need to be rethought. Do a pilot study.

You may need hardware and software. Check out the library, computer and laboratory opening hours. Do you need to book equipment in advance? Have you booked the laboratory equipment for the days after sampling? One day's idyllic rowing across a lake collecting water samples can involve five days' laboratory anal-ysis. In the vacation, that may mean organizing accommodation and time off work.

21.5 Are pilot surveys worthwhile?

Just about every lecturer nags every student to do a pilot survey; about 90 per cent do not bother and about 89 per cent get into a mess as a result. PILOT SURVEYS ARE A GOOD IDEA. They allow you to grab a few samples, trial a questionnaire or run a programme, AND analyse the pilot dataset through all the laboratory, statistical and modelling procedures, putting dots on graphs and numbers into equations. Life is too complicated and the environment much too complex for anyone to get everything right first time. A pilot survey shows there are other angles, other people to consult, suggests other variables to measure, that a screwdriver could be essential and just being in the field, on site, at a computer or whatever, indicates other things you could do to enhance the research. At the very least, go and look at a site before doing oodles of preparation. If you are working overseas, practise your techniques on comparable local sites before leaving. Make sure you have all the right equipment and a complete checklist of the data you need.

At the end of a pilot survey ask yourself:
- Is this the best approach?
- Can I get all the data needed to answer the hypothesis?
- Can the project be done in the time allowed?

Then revise the methodology.

21.6 What's the format and assessment?

A dissertation or thesis is a formal piece of writing. Lecturers expect you to adhere to the 'rules', make the presentation consistent and tidy, and observe the word-length. Ensure that the same typeface is used throughout, that figure titles appear in consistent positions, that there are page numbers on ALL pages, that all the figures and tables are included, and that you double-check the university guidelines and follow them, especially the WORD LENGTH. The general format is roughly along the lines given in Figure 21:1. The word and page lengths are suggestions: there are plenty of high quality dissertations submitted in different combinations. If you have 3000 words for the literature review, it is inefficient to draft 12,000 words.

For assessment information find your departments marking guidelines, or see Figure 21:2. There are many variations; some departments ask for an initial plan that counts for 2–5 credits. The per cent distributions are only a guide; mark distributions vary considerably depending on the nature of the dissertation.

Chapter	Contents		Page/word length
0	Title page, acknowledgements, abstract, table of contents, table of figures.		Keep short
1	Introduction, brief background, research aims, signpost thesis layout.		2–3 6–800
2	Literature review, summary of material relevant to this research. Links to geography, earth or environmental science generally.		4–8 2–3000
3	**Theoretical Thesis**	**Empirical Thesis**	2 – piece of string OR 2000 words
	Methodology, a description of your research approach *or* Discussion and evaluation of first theme/idea/concept.	Methodology, your research process, techniques used, criticism of techniques and evaluation of their accuracy and representativeness. Site information.	
4	Discussion and evaluation of second theme/idea/concept.	Results, with tabulations, graphs, and maps, as required.	More string OR c.1500 words
5	Discussion and evaluation of third theme/idea/concept/ counter-themes. Synthesis of themes and alternatives.	Discursive interpretation of results, this may include further modelling work. Evaluation of accuracy, representativeness, and sensitivity analysis if relevant.	Another piece of string OR c.2000 words
6	Implications for future research. *'What I would have done if I had known at the start what I know now?'* Conclusions.		1 / c.400 1 / c.400
Appendix	Data sets if required. Example copy of questionnaire. Computer program. Sample of interview transcripts.		Minimize

Figure 21:1 Typical dissertation/extended project format

21.7 When do I start writing?

'I'm writing-up' implies writing can be done in one go. Developing ideas and seeing the implications of results takes time. Remember to write as you go. GET SOMETHING ON PAPER EVERY WEEK. Read a couple of things and then draft some paragraphs for the literature review. Add to it when you read the next paper. Writing is part of the research activity: you read a bit, write a bit, think a bit, and the combination of these three activities tells you what you might do

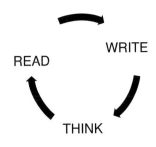

READ WRITE

THINK

next. Similarly write up what you do every time you do something. Capture the action!

You cannot hope to have read all the past literature and to research all aspects of a topic. Much of dissertation management is about drawing a line in the sand, stopping reading, stopping investigating and starting writing. You are aiming to report on what you have read and discovered. Your opinions are based on cited material (*references*) and your results. If you had another year to complete your dissertation, you would still be reporting on partial information. Your examiners want to see that you have tackled a reasonable topic in a relevant manner and drawn sensible conclusions. They do not expect you to model the processes of global warming at the planetary scale, account for the stratigraphy of all Pacific subduction zones, or explain why buses come in threes. Don't get overwhelmed by reading, keep it in balance. Use the checklist in **Try This 4.6** as a guide and remember to balance 20 minutes of online bibliographic searching, with at least a couple of hours of reading.

DO WRITE

THINK

Dissertation Assessment	
Planning Phase	10%
Clarity in formulation of project. Originality in formulation of project. Independent development of project.	
Abstract	5%
Literature Review	15%
Relevance of literature selected? Comprehensive? Critical comments on literature?	
Methodology	20%
Appropriate to topic? Successful in execution? Followed plan and adapted it appropriately?	
Analysis and Interpretation	25%
Planned? Appropriate? Extent to which aims were met? Consciousness of limitations?	
Discussion and Conclusions	15%
Logical and thought through? Sustainability of conclusions? Suggestions for future research?	
Presentation	10%
Quality of figures, tables, maps, and photographs? References? Appendices? Page numbers? Appropriate length?	
Degree of Supervision Required	
Did the student take the initiative? Any illness or personal problems?	
Further Comments	

Figure 21:2 Example dissertation assessment criteria

 TOP TIPS

→ START WRITING at the start: the draft will not be right the first or second time. Ask a tutor how often s/he rewrites a paper before sending it to a journal? 'Lots' is the only answer worth believing.

→ SPELL and GRAMMAR CHECK EVERYTHING.

→ References matter: put them in the document or Endnote from day 1 and double-check your references are properly cited. It is miles easier to do this as you go along. Getting to the last three days and realizing you have no references is like getting to six inches from the summit of Everest and having to go back to base camp for a flag and camera.

→ Proofread what you have written, not what you think you wrote. Make sure there are titles and keys on graphs and maps, scale bars, units with equations and data, and that the title page, abstract, acknowledgements and contents page are in the right format for your department.

→ Proofreading is not easy: get a flatmate to read through for grammar, spelling and general understanding. If your friend understands your 'Environmental Impact Study of Camels in Klatch' then so, probably, will your tutor. You can repay the favour by checking out your mate's dissertation on the 'The Humour of Baldrick'. Leave a few days between writing and proofing – you spot more errors.

→ Abstract: write this last.

21.8 What about safety?

If you break your leg on fieldwork, don't come running to me.

Dissertation research should be good fun but accidents happen; be aware of your responsibilities and safety procedures. (Royal Geographical Society, 2011.) Safety is as important for human geographers in town as for speleologists down a pothole. Each university and department has its own safety policy, you will be given advice, but remember you are responsible for your own actions. Most safety involves common sense and taking sensible precautions. Dissertation fieldwork is usually a solo activity, but do not work alone. Offer 'a day-out' opportunity to friends and relations – two heads are better than one, and the company will keep you cheerful. It is your responsibility to:

✓ Take and keep medication or special foodstuffs with you at all times.

✓ Provide yourself with clothing, boots or wellingtons that are appropriate for your site conditions.

✓ Check weather forecasts.

✓ Leave a plan and map showing where you intend to go with someone who will alert the authorities if you have not returned by a set time. Remember to check in with them when you get back; the police and mountain rescue get very unhappy searching for people tucked up in the local pub reliving the delights of the day.

✓ Take a mobile phone to alert your contact of a change of plan or a problem. In remote areas, taking a first-aid kit, map, torch, compass, spare food and a flask with a hot drink is sensible. Remember, mobiles don't always work in remote areas.

✓ Take no unnecessary risks.

Generally, fieldwork safety advice is aimed at people heading for remote uninhabited areas of Scotland, Iceland or Tunisia, but working in towns and cities can be as hazardous. Take care on unfamiliar roads, and especially in countries where drivers are on the other side of the road. Be courteous at all times with the public. People are often curious. Have a clear, non-confrontational explanation for your activities and explain why you have chosen their area. Think about your clothing and ensure it is appropriate for the locality. If you are not fluent in the local language, ask someone to write a short statement explaining what you are doing that you can show to people. If you are heading for the mountains or glaciers, you may be asked to do a mountain safety course or see the British Mountaineering Council (2011) *Safety on Mountains*, video and booklet. If you are working overseas, see Nash (2011a,b). For developing country work, consult Scheyvens and Storey (2003) and Robson and Willis (1997) which apply equally to undergraduates and postgraduates.

21.9 References and further reading

Dawson C 2009 *Introduction to Research Methods: A Practical Guide for Anyone Undertaking a Research Project*, 4th Edn., How To Books Ltd, Oxford

Flowerdew R and Martin D (Eds.) 2005 *Methods in Human Geography: a guide for students doing a research project*, 2nd Edn., Longman, London

Geoverse 2011 Geoverse e-journal for undergraduate research in geography, http://geoverse.brookes.ac.uk/ Accessed 17 February 2011

Hoskin B, Gill W and Burkill S 2003 Research design for dissertations and projects, in Rogers A and Viles H (Eds.) *The Student's Companion to Geography*, 2nd Edn., Blackwell, Oxford, 197–203

Kitchin R and Tate NJ 2000 *Conducting Research into Human Geography: theory, methodology and practice*, Prentice Hall, Harlow

Knight PG and Parsons T 2004 *How to Do Your Dissertation in Geography and Related Disciplines*, 2nd Edn., Routledge, London

Nash DJ 2000a Doing Independent Overseas Fieldwork 1: practicalities and pitfalls, *Journal of Geography in Higher Education*, Directions, 24, 1, 139–149

Nash DJ 2000b Doing Independent Overseas Fieldwork 2: Getting Funding, *Journal of Geography in Higher Education*, Directions, 24, 3, 425–433

Plymouth Student Scientist 2011 The Plymouth Student Scientist, http://www.theplymouthstudentscientist.org.uk/index.php/pss   Accessed 17 February 2011

Reinvention 2011 Reinvention: a Journal of Undergraduate Research, http://www2.warwick.ac.uk/fac/soc/sociology/rsw/undergrad/cetl/ejournal/ Accessed 17 February 2011

Robson E and Willis K (Eds.) 1997 *Postgraduate Fieldwork in Developing Areas: a rough guide*, 2nd Edn., Monograph 8, Developing Areas Research Group, Royal Geographical Society – Institute of British Geographers, London

Scheyvens R and Storey D 2003 *Development Fieldwork: A Practical Guide*, Sage, London

Swetnam D and Swetnam R 2010 *Writing Your Dissertation: The Bestselling Guide to Planning, Preparing and Presenting First-Class Work*, 3rd Edn., How To Books Ltd, Oxford

Safety: Google fieldwork safety + your university for guidance

British Mountaineering Council 2011 *Safety on Mountains*, video and booklet. Available from http://www.thebmc.co.uk/   Accessed 17 February 2011 Follow Safety and Skills links

Royal Geographical Society (with IBG) 2011 Fieldwork Safety, http://www.geographyteachingtoday.org.uk/fieldwork/   Accessed 17 February 2011

St John Ambulance, St Andrew Ambulance and British Red Cross 2006 *First-Aid Manual*, 8th Edn., Dorling Kindersley, London

Interlinks 1

Starting from the left hand box link across to find six GEES related words. Answers p 304.

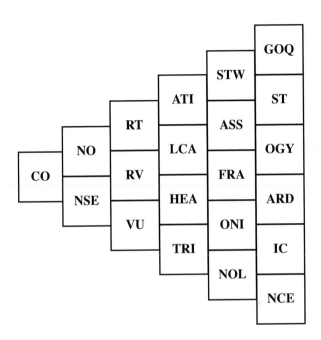

22

Revision skills

I know I knew that once ... What was it?

'I finish five modules this week, have three term papers to hand in on Friday, and the first exam is on Monday. So when do I revise?' Well, the answer is that you don't revise in this style. Revision, in the last-minute 'cramming' style, is really a high school concept. It is part of a Get the Facts → Learn them → Regurgitate in Examination

If you know the technical term, please use it!

process. It is characteristic of surface learning. In most jobs you accumulate knowledge and apply it continuously. University is a transition phase, your deep reading and learning activities mean cramming-style revision disappears.

Rereading and reviewing material is an ongoing process, built into your weekly timetable, because it improves the amount of detail you remember. Normal learning activities reap rewards at exam time because they are also 'revision' activities that reinforce your learning all the time. The following will help:

✓ Get your brain in gear before a lecture by reading last week's notes.
✓ Read and reread all discussion notes.
   Read new things. Talk about them.
✓ Use SQ3R in reading and making connections to other modules (p 60).

> Our best idea was the study buddy group. Explaining something to the others made me understand the points and remember the details.

✓ Think actively around issues (see Chapter 10), ask questions as you research.
✓ Check that you have good arguments (Chapter 11), spot gaps in the evidence and use that information to determine what you read next.
✓ Keep fit, take breaks, eat well.
✓ Timetable 'revision slots' to continue these learning processes.

Tight deadlines are part of university life. Most tutors will be emphatically unsympathetic when you have five essays due on the last day of term or semester and examinations two days later. Sorting out schedules is your problem; an opportunity to exercise your time-management skills. Lecturers know that when they put essay deadlines earlier, to reduce the end of term pile-up, they get more complaints because 'the deadline's too early'.

22.1 Revision aims

Exams test your understanding of interrelated GEES material from coursework, personal research and reading. You will have to overview information on a large number of topics and retain detailed examples and case studies to back it up. Understanding, relevance, analytical ability and expression are listed by Meredeen (1988) as the Holy Grail of examiners keen to donate marks. Reflect on where you can demonstrate these attributes to your nice, kind examiner.

1. **Understanding** – show that you understood the question. Keep answers focused.
2. **Relevance** – eliminate irrelevant examples. Stick to the issues and points the question raises. Don't get sidetracked.
3. **Analytical ability** – aim for a well-reasoned, organized answer. Show that you understand the meaning of the question and can argue your way through points in a logical order.
4. **Expression** – clear and concise writing please, making your ideas and arguments transparent to the marker.

Maps and diagrams help enormously. The length of an answer is no guide to its effectiveness or relevance. A well-structured short answer will get better marks than a long irrelevant answer.

Revision is where you put the puzzle together.

22.2 Get organized, project manage your time

If you take 30 minutes every week for each module, to review and reflect, to think around issues, and decide what to read next, you are revising. It is an ongoing process, and you will be well ahead of the majority! As exams get closer, sort out a work and learning timetable, say for the five weeks before exams. Put every essay, report and presentation deadline on it, add the exam times, lectures, tutorials and all your other commitments, and have a little panic. Then decide that panicking wastes time and sort out a plan to do six essays and revise 12 topics and include social and sporting activities to relax, and block the time in two-hour slots, with breaks to cook, shower and eat (maybe not in that order). Then organize group revision, discussion and sharing opportunities with friends, GET STARTED.

 TOP TIPS

→ Start earlier than you think you need to!

→ Put revision time into your weekly plan and use it to think.

→ Keep a revision record; attempt to allocate equal time to each paper.

→ Speaking an idea aloud or writing it down lodges information in the brain more securely than reading.

Revising is not something you have to do alone. Think with study buddies. Group revision will:

☺ generate comments on ideas and add other people's perceptions to your brain bank;

☺ let two or three brains disentangle theories explaining decentralization, drift or drumlins;

☺ be more fun (less depressing?);

☺ make you feel less anxious, more self-confident;

☺ show where you need extra study and where you are already confidently fluent with the material.

> *I have a wall chart with the weeks and exam dates. Each module has different coloured Post-its. Things to do go on post-its. I move them around, its easy to see where I spend time on each module.*

Don't be put off by someone claiming to know it all because they have read something you have not (no one reads everything); if it's so good, ask for an explanation.

22.3 What do you do now?

What were your thoughts as you left your most recent exams? Ignoring the obvious 'Where's the bar?', jot down a few thoughts and look at **Try This 22.1**.

TRY THIS 22.1 – Post-examination reflections

Write down your thoughts about your last examinations and look at the list below. These are random, post-examination thoughts of rather jaundiced first-year geographers. Tick any points you empathize with, what else concerns you? Make a plan for next time and act on it.

☹ No practice at timed essays since A level; I'd forgotten how after 18 months.

☹ I ran out of time to write.

☹ A lot of the lecture detail seemed unnecessary for the questions.

☹ There were too many specific places and dates to remember.

☹ Lack of revision.

☹ I didn't know what was relevant.

☹ Most of the time I was thinking about going to the bar after.

☹ I needed more direction towards the questions.

☹ You couldn't learn the whole course; I picked the bits that were missed.

☹ There were loads of facts in the lectures, but nothing to apply it to.

☹ I mostly relied on the stuff we did at school and made some things up.

☹ I had all this stuff about Calcutta and he asked about Brazil.

☹ I think the detail in the lectures was confusing and the big handout added another layer of detail that wasn't in the lecture. I didn't really get started on it because there was too much to tackle.

22.4 Good revision practice

The rubrics for papers are usually displayed somewhere in a department or VLE. Find out about the different styles of questions, how many questions you have to answer and what each question is worth. Plan your revision activities to match the pattern of the paper.

> *Making notes from notes is not fun. Recording notes as a podcast and hearing it later was more fun.*

Revise actively

Sitting in a big armchair in a warm room reading old notes or a book will send you to sleep! TRY some of these ideas:

☺ Look at the feedback you have had from coursework, tutorials, the VLE ...

☺ Make summary notes, ideas maps, lists of main points and summaries of cases or examples.

☺ Draw and redraw sections, maps, diagrams and graphs.

☺ Sort out the general principles and learn them.

☺ Look for links, ask questions like:
  • Where does this fit into this course/essay/other modules?
  • Is it a critical idea/detailed example/extra example to support the case?
  • Is this the main idea/an irrelevance?

☺ Write outline answers, then apply the criteria of Understanding, Relevance, Analysis and Expression. Ask yourself 'Does this answer work?', 'Where can it be improved?'

☺ Practise writing an answer in the right time. Sunday morning is a good time! Put all the books away, set the alarm clock and do a past paper in the right time. DON'T PANIC – the next time will be better. Examination writing skills get rusty, they need oiling before the first examination or your first answers will suffer.

☺ Remember you need a break of five minutes every hour or so. Plan to swim, walk or visit the gym; oxygen revitalizes the brain cells (allegedly).

> When did you last use a pen to write for two hours?
>
> Until all exams are typed online, practicing writing with a pen.

☺ Use the Active Reading SQ3R technique (p 60) to condense notes to important points.

☺ Aim to review all your course notes a week before the exams! You will panic less, have time to be ill, and feel more relaxed. (OK, OK, I said AIM!)

☺ Staying up all night to study is not one of your best ideas in life.

Reviewing the first four weeks of a module should, ideally (ho-hum), be completed by weeks six to eight, so that the ideas accumulate in your mind. The rest of the module will progress better because you understand the background and there is time to give attention to topics in the last sessions of the module. Can you be 'ideal' for one module, maybe the really difficult module?

Outline answers

Outline answers are an efficient revision alternative to writing a full essay. Write the introduction and conclusion paragraphs in full, NO CHEATING. For the central section, draft one-sentence summaries for each paragraph, *with references*. Add diagrams, maps and equations in full, practice makes them memorable. Then look carefully at the structure and balance of the essay. Do you need more examples? Is the argument in the most logical order?

Keywords

Examiners tend to use a number of keywords to start or finish questions. Words like Discuss, Evaluate, Assess, Compare and Illustrate. They all require slightly different approaches in the answer. Use the **Try This 22.2** game to replace key words and revise essay answers.

TRY THIS 22.2 – Replacing key words in examination questions

Take a question, any question, from a past examination paper. Do an outline answer structure. Then replace the keywords with one or more of the words below. Each word changes the emphasis of the answer, and the style and presentation of evidence. 'Describe the impact of ...' and 'Evaluate the impact of ...' need different styles of answer. (Keywords and meanings adapted from Rowntree 1988).

Analyse	Describe, examine and criticize all aspects of the question, in detail.
Argue	Make the case using evidence for and against a point of view.
Assess	Weigh up and make a judgement about the extent to which the conditions in the statement are fulfilled.
Comment	Express an opinion on, not necessarily a long one, BUT often used by examiners when they mean Describe, Analyse or Assess (cover your back).
Compare	Examine the similarities and differences between two or more ideas, theories, objects, processes, etc.
Contrast	Point out the differences between ... You could add some similarities.
Criticize	Discuss the supporting and opposing arguments for, make judgements about, show where errors arise in ... Use examples.

Define	Give the precise meaning of ... or, Show clearly the outlines of ...
Describe	Give a detailed account of ...
Discuss	Argue the case for and against the proposition, a detailed answer. Try to develop a definite conclusion or point of view. See Comment!
Evaluate	Appraise, again with supporting and opposing arguments to give a balanced view of ... Look to find the value of ...
Explain	Give a clear, intelligible explanation of ... Needs a detailed but precise answer.
Identify	Pick out the important features of ... and explain why you made this selection.
Illustrate	Make your points with examples or expand on an idea with examples. Generally one detailed example and a number of briefly described, relevant supporting examples make the better argument.
Interpret	Using you own experience, explain what is meant by ...
Justify	Give reasons why ... Show why this is the case ... Argue the case.
List	Make a list of (usually means a short/brief response), notes or bullets may be OK.
Outline	Give a general summary or description showing how elements interrelate.
Prove	Present the evidence that clearly makes an unarguable case.
Relate	Describe or tell a story, see Explain, Compare and Contrast.
Show	Reveal in a logical sequence why ... see Explain.
State	Explain in plain language and in detail the main points.
Summarize	Make a brief statement of the main points, ignore excess detail.
Trace	Explain stage by stage ... A logical sequence answer.
Verify	Show the statement to be true; the expectation is that you will provide the justification to confirm the statement.

The quiz approach

Devising revision quizzes can be effective. You do not need to dream up multiple answers, but could play around with a format as in Figure 22:1. This is a short factual quiz on ozone, based on a journal article by Nyenje *et al.* (2010), but infuriatingly the answers in the second column are out of order. Can you sort them out?

1 Which nutrients are the main causes of eutrophication?	a Increases in phytoplankton primary production; replacement of diatoms by cyanobacteria; water hyacinth blooms; eradication of cichlid fishes; Nile perch population explosion.
2 Where do the nutrients that cause eutrophication principally originate from?	b Population increase, expanding urbanization particularly in informal settlements where wastewater disposal is unmanaged.
3 What were the impacts of eutrophication in Lake Victoria?	c Kampala Uganda, Lagos Nigeria, Accra Ghana, Free Town Sierra Leone, Maputo Mozambique and Nairobi Kenya.
4 What is the average water supply and sanitation coverage in mega-cities across sub-Saharan Africa?	d P (phosphorus) and N (nitrogen).
5 Which cities are cited as experiencing flooding in low-lying slum areas due to stormwater run-off from the urban catchments upslope?	e 70% water supply, <30% sewerage cover.
6 What do the authors conclude is the most significant cause of eutrophication?	f Agricultural areas, urban wastewaters including sewage and from industrial areas.

Figure 22:1 Quiz questions on eutrophication impacts in Nyenje *et al.* (2010). (Answers at the end of the chapter)

The questions in Figure 22:1 cover facts, and are good for MCQ (multiple choice question) revision. Now consider the style of these questions:

1. What does eutrophication mean in practice?
2. What are the implications of not treating wastewater and sewage from urban areas in sub-Saharan Africa?
3. Why do wetlands act as buffers to eutrophication?
4. What are the long-term effects of wastewater disposal to ground water?
5. When and why do some fish species increase their numbers and others become extinct following eutrophication?
6. What are the implications of continued release of organic micro-pollutants to watercourses, for example oestrogens, biocides, flame retardants PFCs and endocrine disruptors?

These questions demand the same initial, factual knowledge as those in Figure 22:1 but a more reasoned and extended response. They are useful for revising short-answer questions and essay paragraphs.

To get into the swing, use **Try This 22.3** and **Try This 22.4**. The first asks for quiz questions based on your notes from lectures and additional reading, and the second uses the same technique with a journal article.

TRY THIS 22.3 – Quiz questions from notes

Pick a set of notes and create a short quiz, 10 questions, in each of the two styles. First, short factual questions, then more extended questions. Write the questions on one side of the page and answers on the reverse, so you can test yourself (without cheating too much).

TRY THIS 22.4 – Quiz questions from a journal

Pick an article from any reading list. Devise a set of five short and five extended questions that explore the topic. Write the questions on one side of the page and answers on the reverse, test yourself next week.

Revision is an opportunity to think about and continue to look for explanations and insights. Revision is therefore a creative process, with ideas gelling and developing as you reread and reconsider, look at new material, make sure you understand definitions, think about timescales and spatial scales. Note-making is an integral part of revision. Some general questions running in your head will encourage a questioning, active approach to revision, see **Try This 22.5**.

TRY THIS 22.5 – Good revision questions

Make a list of questions that could be used in compiling revision questions and quizzes. Some suggestions at pp 304–305, have a go before looking.

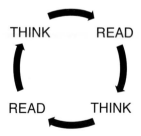

Creating beautiful revision timetables is prevarication at best.
One timetable is enough.

JUST START REVISING!

22.5 References

Meredeen, S 1988 *Study for Survival and Success*, Chapman, London

Nyenje PM, Foppen JW, Uhlenbrook S, Kulabako R and Muwanga A 2010 Eutrophication and nutrient release in urban areas of sub-Saharan Africa – A review, Science of the Total Environment, 408, 3, 447 –455

Rowntree, D. 1988 *Learn How To Study: a guide for students of all ages*, 3rd Edn, Warner Books, London.

Answers to Figure 22:1

1d; 2f; 3a; 4e; 5c; 6b

Wordsearch 3

Find the 20 hazard-related words below. Answers on p 305.

E	G	D	I	C	A	A	I	J	M	U	R	P	R	S
X	K	R	F	W	Y	K	D	F	Z	O	C	I	Z	T
P	Q	A	T	G	R	E	G	N	A	D	S	M	D	O
L	U	Z	U	V	R	N	O	K	T	K	U	A	O	R
O	T	A	O	Q	N	O	C	N	O	O	S	N	O	M
S	H	H	M	O	H	I	A	O	E	E	M	U	L	R
I	E	C	G	P	Z	T	K	E	P	N	K	S	F	M
O	U	G	Y	U	U	P	R	T	I	O	J	T	M	S
N	G	T	Q	L	O	U	O	A	D	L	H	H	P	V
Z	A	O	L	F	S	R	Y	X	E	C	B	R	H	L
M	L	O	I	O	N	E	D	Q	M	Y	F	E	B	F
I	P	K	P	A	N	D	E	M	I	C	H	A	P	I
F	M	X	D	I	P	F	E	O	C	G	N	T	Y	H
V	E	O	O	Q	C	H	Y	K	C	S	R	N	H	R

Examinations

Examinations should not be a journey into hitherto uncharted regions.

Be practical: relax, you did all that reflection and reviewing; the examinations will be most agreeable. Check and double-check the examination timetable and room locations – they can change. Know where you are going, plan to arrive 20 minutes early to find your seat number, visit the loo and relax. Check the student handbook for what to do if delayed. Make sure you have your ID, pen, spare pens, calculator ... whatever is required. For online exams you need your ID and passwords. Read the information on the exam paper

> Double-check all examination dates and locations

(rubric), and take heed of everything the invigilator says. If there is a problem with the paper or computer tell an invigilator at once, so the lecturers can be consulted.

Find out in advance how the paper is structured and use the time in proportion. For example, on a two-hour paper that has six short questions and one essay there will probably be 50 per cent for the essay and 50 per cent split equally between the six short questions. That's one hour for the essay and nine minutes for each short question, leaving six minutes to read the paper and plan the answers. Answer the number of questions required, *no more* and preferably *no less*. Leave time to do justice to each question. Don't leave your potentially best answer until the end. Equally, don't spend so much time on it that you skimp on the rest. Make sure there are no questions on other pages.

Should you be seized with anxiety and your brain freezes over, use the 'free-association' brainstorming approach.

> Read the module handbook.
> What are the learning outcomes?
> Have you achieved them?
> Handbooks often include advice on exams, some contain past question papers and model answers.

Write out the question and then look at each word in turn, scribbling down the first words that occur to you, authors' names, examples, related words ... This should generate calm and facts. Plan from the spider diagram you have created.

23.1 Examination essays

The minimalist advice here is to reread the advice on argument (Chapter 11), revision (Chapter 22) and writing essays (Chapter 17) and to write fast. Essays allow you to develop lines of argument, draw in diverse ideas and demonstrate your skills in argument, analysis, synthesis, evaluation and written communication. Remember to keep the academic content – geology, management, economics, petrology, biochemistry, hydrology, legal, geography, science, philosophy – high, use evidence (lots of *references* and *case examples*) to support your arguments and cite the *references*.

> **Essay Plan Lite**
> - Introduction: background, context, framework
> - Ideas, examples, references
> - Ideas, examples, references
> - Ideas, examples, references
> - Interpretation and discussion
> - Conclusions

Good answers communicate enthusiasm for the material, avoiding boring regurgitation. Persuade the reader to agree with your ideas.

⇨ TOP TIPS

→ If all questions look impossible, choose one where you have examples to quote, or the longest question. Long questions give more clues to structure an answer. 'Discuss the role of pressure, mineralogical crystal structure, temperature and other variables on radioactive decay in three contrasting rocks' is a question with loads of clues and parts to answer, whereas 'Discuss radioactive decay processes' is essentially the same question. It could be answered with the same information, but will be a minefield if you do not impose structure and facts.

→ Always plan your answer. Underline key words in the question, like <u>Discuss</u>, or <u>Compare and contrast</u>, and note the spatial and temporal scales to cover (Figure 23:1). Don't restrict examples to one place or country if the question asks about global or international issues. Do a quick list, mental map or spider diagram of the main points, and note ideas for the introduction and conclusion. Then rank the points to get a batting order for the sections.

> *It was two years before I realized you were expected to put references in exam answers, with dates.*

→ On a three-question paper, plan questions 2 and 3 before writing the answer to question 2. Your brain will run in background mode on ideas for question 3 as you write the second answer.

→ Watch the time. Leave a couple of minutes at the end to check each answer. Amend spelling, add extra points, references, and titles, scales and units to diagrams and maps.

→ Organize points, one per paragraph, in a logical order. Set the scene in your first paragraph and signpost the layout of the answer.

→ Be specific. For example, rather than 'Early on ...', give the date, or 'In the Jurassic era ...'. 'In UK towns we see. ..' is better as 'In Winchester, Reading and Shrewsbury we see ...'

→ Put an interesting point in the final paragraph.

The comments in Figure 23:2 suggest that examiners only give marks when you explain what you know. Clarify terms, answer all parts of the question, include many examples, remember your discipline. Geographers need to include geographical cases and examples. Unless you relate theory to the real world, you are not going to hit the high marks. Similarly for geologists, you might give a very good technical answer about a chemical process but for high marks explain where and why this is relevant for geology.

Use Technical Language
- Avoid contractions: it's, let's, doesn't, wouldn't ...
- Put in units, formulae, equations
- Add diagrams, graphs, data
- Avoid colloquialisms (see p 184).

If you are going to cross things out, do it tidily.

1:1	Hill slope	mm	cm^2
1:50	Channel	m	m^2
1:250	River corridor	100m	hectare
1:10,000	Floodplain	km	km^2
1:50,000	Catchment		
1:1,000,000			
Truro	Parish Council	By-laws	
Cornwall	Local Authority	National Law	
South-west England	Regional Authority	International Law	
United Kingdom	National Legislature		
European Union	International Groups		

Second	Household	Eon	Genetic diversity
Minute	Ward	Era	Species diversity
Day	Town	Period	Ecological diversity
Year	County	Epoch	
Decade	Region	Age	
Century	Nation	Chron	
Millennium			

Figure 23:1 Spatial, temporal and other scales

Does not address the question	25%
Swarming with factual errors	35%
Fine answer to a question about coasts, pity the question was about karst	38%
Very weak effort – no attempt to explain examples or definitions	42%
Somewhat confused but some relevant points	42%
Good background but little analysis on the question set	44%
A scrappy answer with some good points	45%
Started fine then got repetitive	49%
Decent attempt, some examples but not focused on the question	50%
Not sufficient explanation/evaluation, some ideas listed at the end but these needed developing to raise the mark	52%
Reasonable answer but misses out the crucial element of ... No equations, no science	54%
Has clearly done some reading but fails to write down the basics	54%
Reasonable effort as far as it goes, but does not define/explain technical terms – no examples, no references	55%
Accurate but generally descriptive, never really got to the 'evaluate' section	55%
Needs to learn about paragraphs, then fill them with organized content	55%
Reasonable effort – covers many relevant points, structure adequate but needs to organize points in shorter paragraphs	57%
Too much description – not enough argument	58%
Good – but needs to be more focused to key points	58%
Good introduction and well argued throughout, although with no evidence of reading. Good use of examples but unfortunately only from class material	64%

Has done some reading and thinking, lacks depth and references	65%
Quite good – a general answer, omits to define land-use but lots of non-lecture examples	64%
Excellent, comprehensive evaluation with new material	75%
Outstanding, original research as well as new reading and some original thinking	85%

Figure 23:2 Comments on examination essays

Take-away or take-home examinations

Essays written outside an examination room need revision and preparation. Get the notes together in advance, make sure you have the library resources you require before the paper is published to avoid being trampled underfoot in the library sprint. Get your brain in gear by thinking of possible essay topics. Do not leave it all to the last minute.

23.2 Short-answer questions

Short-answer questions search for evidence of understanding through factual, knowledge-based answers and the ability to reason and draw inferences. For short answers, a reasoned, paragraph answer is required. If the question starts 'Give four explanations for ...' ensure you make four, fact-rich points. Take care to answer the question that is set! Add diagrams, graphs, equations and formulae as appropriate.

23.3 MCQs

Multiple choice questions (MCQs) test a wide range of topics in a short time. Modules with MCQs usually include practice sessions: be there. Make sure you understand the style of questions and the rules. MCQs may be used for revision, in a module test where the marks do not count, or as a part of module assessment where the marks matter. Lecturers design class tests very carefully to let you appreciate what you already understand, and where more research, thinking and revision will be helpful. In a final assessment, watch the rules.

> Did you know all the MCQ questions and answers are scrambled in the test? We do the same 50 questions, but they come up randomly. Trying to compare afterwards was hilarious. Matt wanted to know what we thought was the answer to question 4, but we all had a different question 4.

Look carefully at the instructions on MCQ papers. The instructions will remind you of the rules, such as:

There is/is not negative marking. (With negative marking you lose marks for getting it wrong.) *One or more answers may be correct, select all the correct answers.* (These can get you 100 marks on a paper with 60 questions).

General advice says to shoot through the paper answering all the questions you can do easily, and then go back to tackle the rest (but general advice does not suit everyone). Questions come in a range of types:

The 'Trivial Pursuit' factual style

These test recall of facts and understanding of theories, usually a small proportion of the questions:
1. CPRE stands for:
    a) Centre for the Promotion of Regional Excellence
    b) Capital Protected Rate of Exchange
    c) Council for the Protection of Rural England
    d) Center Parcs Rafting Exercise
2. Darcy's Law is usually expressed as:
    a) Q = CIA
    b) Q = KIA
    c) Q = K(h/l)A
    d) Q = CID
In this last example there are two correct answers, B and C, and both should be indicated for full marks.
3. Which of the following was not a super continent?
    a) Pangaea
    b) Laurasia
    c) Davros Skaro
    d) Kenorland

Reasoning and application style

Reasoning from previous knowledge gives rise to questions like:
    Which of the following sequences correctly ranks air pollution emissions in the UK in 2010 (lowest emission first)?
a) Nitrogen oxides, Sulphur dioxide, Black smoke, Carbon monoxide
b) Black smoke, Nitrogen oxides, Carbon monoxide, Sulphur dioxide
c) Black smoke, Nitrogen oxides, Sulphur dioxide, Carbon monoxide
d) Black smoke, Sulphur dioxide, Carbon monoxide, Nitrogen oxides
Some questions give a paragraph of information and possible responses. You apply theories or knowledge to choose the right response or combination of responses. Some tests combine information from more than one course, as here, where a

second-year soil analysis paper requires information from first- and second-year statistics modules:

A soil survey of 25 fields yielded measurements of NO_3, NH_4, P, K, particle size, OM, and bulk density, crop yield (tonnes per acre). The fields were planted with 4 crops, (8 of barley, 8 of broccoli, 2 of sugar beet and 7 of broad beans).

1. Which statistical tests might be used to analyse the data set comparing production of all 4 crops? Mark all the correct answers:
   a) Mann Whitney U, b) Student's t-test, c) Correlation, d) Multiple regression, e) One-way Analysis of Variance, f) Histograms, g) Chi-square, h) Nearest Neighbour analysis, i) Kruksal-Wallis test, j) Two-way Analysis of Variance.
2. Which statistical tests should not be used to compare broccoli and sugar beet yields? Mark all those not correct:
   a) Mann Whitney U, b) Correlation, c) One-way Analysis of Variance, d) Histograms, e) Chi-square, f) Nearest Neighbour analysis, g) Kruksal-Wallis test, h) Two-way Analysis of Variance, i) Student's t-test, j) Multiple regression.

Data response style

A combination of maps, diagrams, data matrices and written material are presented, and a series of questions explore possible geographical, environmental and geological interpretations.

Appropriate revision techniques for MCQ papers include deriving quiz questions. Revisit **Try This 22.3** and **22.4**. Generating questions forces you to concentrate on details, and is a great way to revise.

23.4 Laboratory module examinations

Some laboratory courses have examinations that test the applications of laboratory work. Revise with past papers, or make up your own questions. Try variations on the following questions:

- *If soils experiment … was rerun with soil samples from Exmoor/Tunisia/Spitsbergen how would you expect the results to change?* Essentially a 'what will happen if the samples come from different environments' question. They are designed to test understanding of controlling principles.
- *The flume tests were rerun with flows of … m.s⁻¹. How would you anticipate the sediment erosion and deposition patterns would alter from those observed in the experiments?* This is another 'what if we change the parameters' question that is asking for logical speculation on outcomes.
- *After Experiment 12 was completed the … meter was found to have an error of 15%. Explain how this information affects the interpretation and how it can be*

accounted for in the calculations. A 'how can you cope with instrument and operator error' question.

- *Explain where errors can arise in ... [water sampling, pollen counting] and the impact they have on data analysis and interpretation.* A general question about data and analysis errors.
- *Describe the safety implications involved in ..., and explain how they can be minimized.*
- *Explain how and why instruments to measure ... should be calibrated before field use.*
- *Outline the field sampling and laboratory tests you would use to explore ... How accurate would you expect your results to be?* Explores the approaches and tests you would use, and their relative accuracy.

23.5 Oral examinations

Vivas

Vivas are oral examinations. They are an increasingly rare but in some degree schemes, in some universities they are part of a module. Vivas are an exciting opportunity to talk about a topic or project. Think of it as a presentation without a presentation, you move straight to the Q&A session. This helpful skill should be more widely practised, since in the workplace you are much more likely to 'explain an idea or project' than to write about it. Don't panic!

After all the assessments and examinations, when all the marks are sorted out, some people are 'borderline specialists', people so close to the boundary that a viva is given. A viva may also be given to candidates who have had special 'circumstances', such as severe illness, during their studies. The viva is usually run by the external examiner, a delightful professor or senior lecturer from another university. DO NOT PANIC. In all the departments I know, vivas can only raise candidate's marks. You have what you have, things can only GET BETTER. Remember all your previous revision and think round some answers to the questions below. Get a good night's sleep, avoid unusual stimulants (always sound advice) and wear something tidy. Examiners are interested in your brain not your wardrobe. Celebrate later.

The dissertation viva

You are the expert, you did the research and wrote it up. The best advice is: read your thesis, think (and make notes) about what you know now that you didn't know when you started. Be ready for easy starter questions:

- Can you please start by summarizing what you did and what the results mean for ... geology/tectonics/urban studies ... ?

- What were the three most challenging things and how did you handle them?
- What should a student working on this topic next year do to make the most of what you have done and develop the ideas/understanding ... ?
- What are the two main strengths of your dissertation research?
- Outline any weaknesses in the research.
- Explain why you adopted your research strategy.
- What were the main issues you addressed?
- Since you started on this topic I see you did a module on ... How might you have adapted your research having done this module?
- Explain how your results relate to hydrology/cultural geography/petrology/limnology/geology/sustainability/environmental chemistry ... generally?
- Obviously in a student dissertation there is limited time for data acquisition. How might you have expanded the data collection process?
- Could you talk about sources of error in the data please?

The general viva

The examiner will ask questions around the modules and projects you have undertaken. Anticipate questions similar to those above.

In both dissertation and general vivas keep answers to the point and keep up the technical content. Any examiner recognizes waffle. External examiners have supervised, marked and moderated thousands of papers. They know every dissertation has strengths and weaknesses, they are impressed by people who have done the research and realize there were other things to do, other ways to tackle the issue, other techniques, more data to collect ... so tell him/her that you know that too.

Use your 'GEES-speak' skills. Asked about the 'urban village concept' you could say: 'It was an idea that started in America in the 1960s, and a number of people wrote about it. Then urban villages declined as economic pressures changed. People moved out to the suburbs. More recently there has been reinvestment in the city centre, leading to an increase in village-style communities within city centres. In the UK, the idea of urban villages is beginning to catch on with planners'. It is a general answer but in the 2.2 classification. You could say: 'Gans described the urban village concept in 1962. He looked at the way Italian–American immigrants in Boston lived an essentially rural lifestyle within the city, preserving traditional social structures and, in some cases, their language. The break-up of these core communities followed the transfer of industry and populations to the periphery of the city. However, in the last ten years, the regeneration of inner-city areas, gentrification of property, an increase in service sector jobs in the city centre and a desire to live in the centre, close to work and leisure opportunities, has revitalized city centres, and urban villages are re-emerging, albeit in a different form. In the UK in 1997, the Urban Villages Forum published a description of urban villages and explained why they can be useful in planning British cities'. A longer and more geographical answer, the references are there, Boston is an example and there are geography terms like core, periphery, social structures and regeneration. This answer has the evidence and should get a First. So think a bit and answer with evidence, and examples, and references, just as in an essay.

After that I think you cannot do much more preparation. The examiner can ask about any paper and question, about the degree as a whole and your own subjects in general. If you missed a paper or question, or had a nightmare with a particular paper, think around answers to the questions BUT the odds are they will not be mentioned.

 TOP TIPS

After every exam, aim to review your answers as soon as you can. What did you learn from the exam process? University exams are a little different to school ones; a spot of reflection on your revision and exam technique might be useful before next time. Questions that might occur to you include:

→ Had I done enough revision? The answer is usually *no*. Ask '*Where could I have squeezed in a little more revision?*'

→ Was there enough detail and evidence in each answer to make an excellent case?

→ What revision activities would be useful next time?

23.6 Feedback

There are many ways students get feedback on their exam performance. Find out what your School does and get involved. There may be:
✓ A day when all examiners are available with all the papers. You go through what you did well and where you can improve.
✓ Your tutor asks you to come for a chat.
✓ Good answers posted on the VLE with comments on why this was an effective answer.
✓ A podcast of the markers comments for individuals or for the whole class.
✓ Comments sent to your email account.
The trick is to use the feedback effectively. You cannot change the mark, the module is over and maybe you will never again read or think about the topic, but turning up and talking helps you review the process of revision and writing exam answers that will help you next time.

23.7 References and further reading

Search for MCQ examples for your subject. Some of high school ones make a nice introduction to revision.

Hay I 2002 Coping with exams: dealing with the cruel and unusual, in Rogers A and Viles H, (Eds), *The Student's Companion to Geography*, 2nd Edn., Blackwell, Oxford, 190–196

Knight PG and Parsons T 2003 *How to do your essays, exams and coursework in geography and related disciplines*, 2nd Edn., Routledge, London

Levin P 2004 *Sail through exams*, Open University Press, Maidenhead

Loughborough University 2011 Revision and exam skills, http://www.lboro.ac.uk/library/skills/Advice/Revision%26Exam.pdf   Accessed 17 February 2011

Open University 2011 Revising, exams and assessment, http://www.open.ac.uk/skillsforstudy/revising-exams-and-assessment.php   Accessed 17 February 2011

Wordsearch 4

Below are 24 familiar GEES-minded terms to find. Answers on p 305.

```
E E H T S R A K O B L D N N C
N B M E O I N C A R U O O A A
E D S S M M E S D O I B S O R
T E I A E A B L T R A C R B
I L H L N L T C A A B G I S O
S C P I T E D I C I T S E P N
A Y R N L L D O T T A M W E A
R C O I M A R B E E R U I C T
A E M T R D T E D E I I L I E
P R A Y Y G A B B R O V D E N
L I T H O S P H E R E U L S O
N D E R E G N A D N E L I I B
E O M E L O R E M U F L F B R
D R O J F R E A K B A A E E A
A L E N E R G Y L H F R C U C
```

24

Field-classes and fieldwork

*A geopolitical boundary is an imaginary line between two nations,
separating the imaginary rights of one, from the imaginary
rights of the other.*

Fieldtrips are one of the great bonuses of GEES degrees. The occasional student confuses a field-class with a fantastic holiday with mates, but the staff have a well-planned learning agenda for you. Costs are an issue, be clear about them at open days, and briefings. GEES departments donate a considerable proportion of their budgets to fund the teaching of fieldwork and field mapping exercises because they are convinced it:

☺ develops observation and research skills;

☺ gives a realistic insight into research activities;

☺ allows hands-on, practical experience of field techniques;

☺ encourages individuals to take responsibility for research decisions;

☺ develops group and interpersonal skills;

☺ encourages students to talk through research issues with staff in small group research work;

☺ allows students to experience some unfamiliar environments.

Students usually pay for travel, some accommodation and food costs. Your department will provide detailed briefings including costs at open days, in handbooks and online, as well as safety information for each trip.

This very short chapter aims to be helpful to pre-university readers too. Fieldwork is a short experience. Staff aim to give a 'taste' of research activities. The speed of a field class may mean your technical achievements are limited. At the end of a field visit list the skills you have used and improved, and add them to your reflective log or CV. Skills acquired through fieldwork include organization, independence, teamwork, measurement, description, safety, use of maps and statistics. Also communication and presentation skills through podcasts, videocasts, posters, maps, field notebooks, reports, negotiation and discussion.

24.1 When is it?

Residential field and mapping courses are usually booked a year in advance to get good rates for travel and accommodation. Find out the dates and plan accordingly.

First-year field trip dates and costs should be advertised in pre-course literature. Do not agree to skiing holidays, rugby trips or family silver wedding celebrations during this period. Field trips are usually a whole or part of a module, therefore the marks matter.

24.2 Time management

Generally these are easy modules to manage because your lecturers organize you. Expect something like: breakfast at 8, leave at 9, exciting field activities 10–5, dinner at 6, analysis, discussion and report from 7.15, hand in written report at 9 the next day. Field-classes prove you can get up and be ready to work at 9 every day for a week, despite having enjoyed the company of colleagues for much of the night. Parents are prone to wonder why you need three days' sleep on return from field-class. Tell them something comforting about sad, sadistic, but fit, lecturers that made you walk 9 miles each day, AND collect data, AND use computers, AND write it all up before bed AND your room-mate snored – or something similar.

24.3 Preparation

Practical

Six weeks before a field-class check out your passport, injections, holes in wellies, and arrange to borrow rucksacks, wet-weather gear, clipboards and pens and pencils that keep writing no matter how hard it rains. Ask people who went last year what you really need to take, BUT just because last year there was full sun on Denbigh Moor does not mean the temperature can be guaranteed to rise above arctic-survival-bag point this year. Unreasonably, GEES staff want to view the landscape from the top of the hill, and the hill is rarely equipped with a road or a path. Consider borrowing boots. Geologists should prepare to head into quarries and cuttings that are muddy and the other side of the hedge, nettles or gorse bushes. By week zero you should be ready.

Academic

Where are you going? It makes sense to look at a map and find out about the place, whether it is Bangor or Bangalore. Some departments have pre-departure exercises and briefings, others rather assume that as an intelligent 18+ person you will know that it may snow in the Mediterranean in January and that you need euros to buy drinks in Greece. Ask last year's group about what to expect and to take. Reminiscing about field-classes over a few pints is normal GEES business.

There may be virtual fieldwork resources to view. Information from other classes that have visited the area. Search online using open forum + fieldwork + your location and see what pops up. Baines (2011) has a brilliant virtual guidebooks site that is worth a visit before any trip to the USA, for fieldwork or a holiday. Don Bain is a retired Geography lecturer from Berkley, California who has created a massive resource of panoramas that you allow you to see around the landscape in 360 degrees. YouTube also provides some fieldwork insights.

24.4 Safety

Someone will alert you to fieldwork safety issues. Most departments have safety codes, some will ask you to sign to confirm that you are aware of safety issues, insurance and your responsibilities. Generally you will be expected to:

✓ co-operate with the staff about safety and behaviour in the field;

✓ act in such a way that neither you, nor anyone else in the group, is put at risk;

✓ not wander off or do something hazardous;

✓ make staff aware in advance if you have a medical condition such as allergies, asthma or diabetes;

✓ know what to do if something goes wrong, such as becoming detached from the group, missing the bus or becoming disoriented if the weather deteriorates and visibility is lost;

✓ have appropriate clothing.

Lecturers have completed risk-assessment documentation, planned for problems and gained access permissions. They are usually first aiders. They will check the weather and adjust activities accordingly.

In rough terrain, field paths, hills, anywhere out of town, walking boots are safest. For river work, or on peat bogs, wellingtons with soles that grip are useful. Aim for lots of thin layers: T-shirt, shirt, sweater and fleece under a cagoule is ideal. Layers can be removed, or added, during the day. A woolly hat and gloves are essential for fieldwork in the UK from September to May, and ideally waterproof trousers and a bright cagoule with hood, to go over an anorak or fleece. Hard hats will be supplied when needed. Good field gear is expensive; can you borrow clothes for your field-class?

Safety is mostly applied common sense. As a general rule of thumb, 'if you wouldn't do it with Grandma or a baby on a Sunday walk' then probably it is a bad idea. On the bad ideas list are: standing in the middle of roads to sketch buildings; trespassing on private property; entering caves, mine workings, working quarries, derelict buildings; or any moving water that is more than 5cm deep. Stick to public footpaths

and rights of way. Wear hard hats in quarries, under cliffs and on steep slopes, where rock or debris falls are a hazard. Do not climb on cliffs or rock faces. Take care on foreshores, rocky beaches and keep an eye on the tide. Do not light fires and extinguish cigarettes very carefully.

Follow your lecturers' advice. Soil and water can carry bacteria, viruses and other pathogens that cause disease. Use appropriate sampling methods, gloves when taking water samples, for example, and ensure that you wipe your hands before eating. Everyone on fieldwork should be sensible, polite, considerate of others, and take special care not to damage anything, from hotels to field walls, or to leave gates open.

Take care and have fun, most field-classes are very well planned and free of incidents.

And while away, be environmentally friendly. Use recycled toilet paper.

24.6 References

Bain D 2011 Don Bain's Virtual Guidebooks, http://virtualguidebooks.com/ Accessed 17 February 2011

See also the Safety references in Chapter 21, p 215.

Interlinks 2

Starting from the left hand box link across to find six GEES-related words. Answers on p 305.

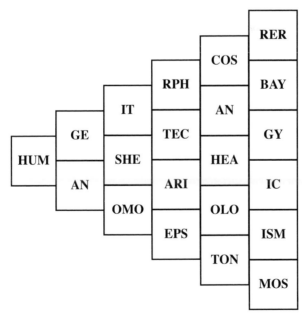

Presenting field sketches, sketch maps, and data

A graphics program or sketch will move mountains for you.

Maps and diagrams save words; they are good value, especially in exam answers where time is short. Plotting, sketching or mapping GEES information exemplifies and clarifies relationships.

Everyone doing science or map work at school learns about labelling, scales, legends, keys and titles *because every map and diagram must have them*. Lecturers cannot give full marks if they are missing. This is not because life is grossly unfair, which it is, but because geographers, environmental and earth scientists always label all maps and diagrams correctly! The skill here is to give careful, consistent attention to detail and completeness. Don't forget to acknowledge sources with references in figure titles, as in 'Figure x ... title ... (after Penman and Wipe 2020)'. See also Chapter 18.3, p. 192 for advice on presenting field notebooks.

Drawing happens in examinations: 'Provide an annotated plan of a new retail park, with brief justification for the features included', 'Draw a geologic map to illustrate sinistral and dextral wrench faults', 'Sketch the micro-depositional forms you find in alluvial channels'. Check past exam papers.

There are deliberate mistakes on the figures in this chapter, please exercise your critical skills and improve each one.

25.1 Field or panorama sketches

Typically, you will find yourself walking happily in Arran, Abingdon or Algeria when some sad academic says brightly, 'Please do a field sketch of the lithological/ urban/cultural /geomorphological /biogeographical/... features'. Everyone mutters about not being art students. GEES lecturers have heard this before and will be deeply unimpressed. A rough sketch can convey considerable geological, environmental and geographical evidence making sketches invaluable in reports and essays. Good lecturers sketch too so you can see what to do. Figure 25:1 is not particularly good, being part sketch and part cross-sectional plan. There is considerable room for artistic improvement, but it will get good marks because it picks out the evidence for slope movement through specific examples. It also shows a 'wider view' than a camera. Five photos would be required to show all these points. The sketch includes evidence the camera cannot 'see' behind cars and trees.

Great field sketches take practice, but not much. The trick is to pick out the two or three lines that anchor the sketch, usually the horizon, maybe a building or two, and then place the features you want to illustrate. Geologists focusing on rocks, faults and surfaces find that adding a building or a tree helps too (See Davidson 2008, and Williams 2010).

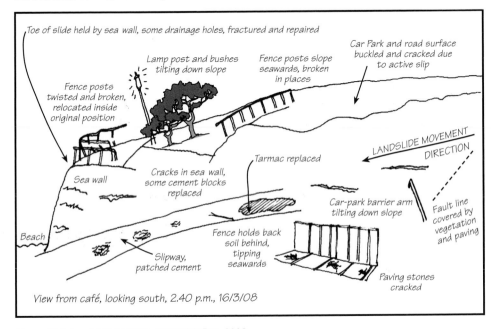

Figure 25:1 Landslide impacts, Runswick Bay, 2008

⟹ TOP TIPS

→ Draw the features 'that matter for this exercise only'. Attempting to draw the whole landscape is a waste of time (unless you are Constable reincarnated). Keep it simple.

→ Helpful tools include a firm board, paper, 2B and 4B pencils, pen, eraser and, perhaps, binoculars.

→ Locate a couple of 'landmarks' to anchor the sketch, then simplify.

→ Show general outlines for buildings or slopes, omit details.

→ Shading gives a feeling of depth and distance – turn the pencil sideways. Use thicker, darker lines for close objects and finer lighter lines for those in the distance.

→ Everyone exaggerates the height of objects like hills and buildings to start with. Try a second sketch reducing the heights by half to two thirds and compare the result. Shadows clutter a sketch; they are a transient element, a function of the weather and time of the visit, and best omitted.

An artist sketching a landscape covers his work with notes on colour, shading and lighting effects. Good field sketches are covered with notes making relevant hydro-logical/geophysical/cultural … points. Notes get marks.

First-time sketchers can find it helpful to start with a sky sketch! Skies are gener-ally empty, clouds have simple shapes (Figure 25:2). The only landscape element required is the horizon. Try it from your window now! Minimum labelling involves naming the features, in this case, the types of clouds, then add information about cloud elevation and whether there will be rain soon. Look at Figure 25:2, what other information might be added? Be critical.

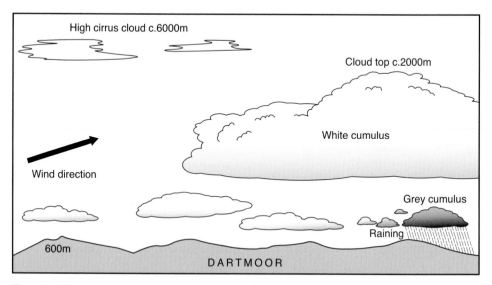

Figure 25:2 Cloud sketch, view from SX417599, clear day, 11.30 a.m. 20 September 2010

Figure 25:3 shows four sketches of the same site, drawn to illustrate four differ-ent aspects of the landscape. Pick out the geological/environmental/geographical features you want and ignore the rest. Excellent answers would add a small map

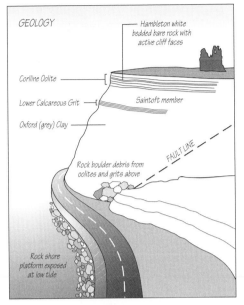

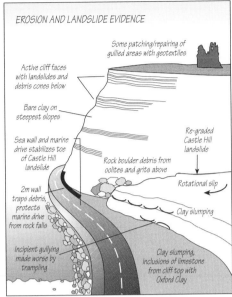

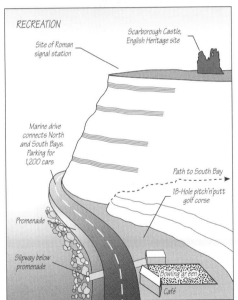

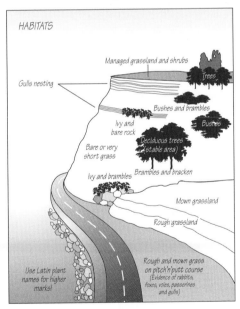

Figure 25:3 Sketches of Castle Hill, Scarborough, UK

to locate the view, the GPS or grid reference of the sketching point and a compass bearing. Make sure your notes are legible.

Figure 25:3 started as sketches which were scanned into a graphics program and the shading added. They may look more 'professional', but the hand-drawn version is perfect for field reports and dissertations.

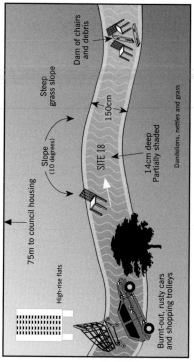

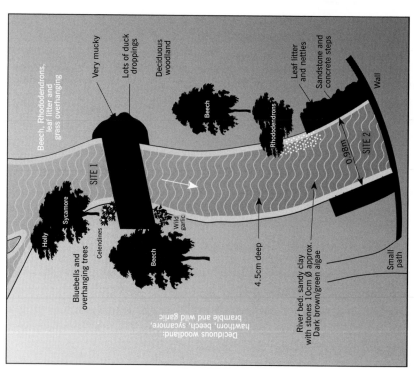

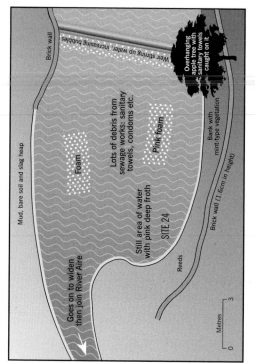

Figure 25:4 Sketch plans of three stream reaches. What is missing?

25.2 Field plans and maps

Figure 25:4 shows sketch plans of urban stream reaches. They make multiple points about flow, vegetation and debris. An indication of SCALE is vital: this is a small tributary, not the Mississippi. One diagram has a scale bar, on the other two the width of the stream is shown. In the final version, replace the plant names with their Latin names (see p 280). Good examples of field plans and sketches can be found in *RSPB et al.* (1994).

Sketch maps highlight specific geographical and geological evidence. Either sketch freehand or use a published map as a base, and annotate copiously. Unless there is a good reason, put north at the top of the map, add a direction arrow and indication of scale. Simplification is the key, do not attempt to be comprehensive, convey specific information. 'Anchor' the map with landmarks that will be there if you revisit in ten years later: river, railway, power station.

It is logical to use conventional signs from Ordnance Survey and geology maps (Ernst 2006). For a black and white sketch some adaption will be necessary, for example, to distinguish a road from a river. Symbols used in geomorphological mapping (Cooke and Doornkamp 1990, Hayden 2009, Universitat Salzburg 2010) are also useful.

25.3 Tables, graphs or data

Manipulating data to create tables, histograms and graphs is straightforward; inserted into reports essays and dissertations they look professional and save words. They are not, of course, a substitute for thoughtful analysis. Look very critically at your own and published diagrams, most diagrams can be improved. Tables of data are inherently boring. Think about how the data can be displayed more interestingly to support your argument.

 TOP TIPS

Start by clarifying information for the reader. Use this checklist:

→ Is the data correct? A very high or low data point can indicate a data entry error. Double check, look back at field notebooks and spreadsheets. Take plenty of time because a data error can totally alter your interpretation of the results.

→ Do graphs have full and explanatory axes labels, units, and titles?

→ Does the presentation design make the information difficult to read and understand?

→ Is the referencing complete? Does the citation acknowledge the source of the data?

→ Is it clear whether the data is primary or from another source?

→ Are there error bars on graphs and histograms?

→ Do mean, median or modal plots show the data range indicating the precision of the data?

→ Do regression plots include r values?

→ Do regression lines fit the data? Check there is a balanced distribution of points above and below the line.

→ Where there are multiple graphs on a page are the scales on each graph the same, enabling visual comparisons? This problem arises when computer packages scale the y-axis to fit the range of the data.

Computer packages will create and colour graphs, bar charts, pie charts and all sorts of diagrams. Don't let the IT get in the way of your message. Aim for illustrations that are clear, legible, interesting *and* convey your message accurately.

25.8 References and further reading

Cooke RU and Doornkamp JC 1990 *Geomorphology in Environmental Management*, 2nd Edn., Clarendon Press, Oxford

Davidson R 2008 Field Sketching, http://www.map-reading.com/appenda.php Accessed 17 February 2011

Ernst WG 2006 Geologic mapping – where the rubber meets the road, in Manduca CA and Mogk DW Earth and Mind: how geologists think and learn about the earth, *The Geological Society of America* Special Paper 413, Boulder, Colorado, 13–28

Hayden RS 2009 Mapping, Geomorphology from Space, NASA, http://disc.sci. gsfc.nasa.gov/geomorphology/GEO_11   Accessed 17 February 2011

RSPB, NRA and RSNC 1994 *The New Rivers and Wildlife Handbook*, The Royal Society for the Protection of Birds, Sandy, Bedfordshire

Universitat Salzburg 2010 Geomorphic Symbols, geomorphology, at http://www.geomorphology.at/index.php?option=com_content&task=view&id=133&Itemid=176   Accessed 17 February 2011

Williams M 2010 Field sketches and how to draw them, University of Liverpool, http://pcwww.liv.ac.uk/~hiatus/Resources/field%20sketches/Index%20field%20sketches.htm   Accessed 17 February 2011

Words in Geo-words 4

Set 8 minutes on the timer (cooker, mobile ...) and see how many words you can make from:

B A S A L T

Answers on p 305.

26

Posters and stands

Aim for short, clear, relevant and short.

Walk around any university department and you will find posters on the notice boards. GEES researchers share their ideas and results at conferences through posters, and they are used frequently in the workplace. A display stand at an event usually includes one or more posters together with materials for participants to take away. The very best posters and stands communicate messages clearly and concisely.

26.1 Posters and e-posters

A poster is not an essay in very small font on one page.

Posters have limited space, forcing presenters to concentrate on the essential elements expressed creatively through brief, concise statements and explanations. There is no space for waffle. GEES departments display staff and student posters on corridors, office and laboratory walls. Take a critical look at them. How effectively is the 'message' communicated? Are the main points readable from 2m? Are you enticed into going closer and reading the detail? Do you like the colour combinations? Is there too much or too little material? Is there a good mix of images and written information?

Check out Macrae *et al.* (2010) for a good academic example. The Macaulay Institute (2010) which has a set of five soils and four vegetation posters aimed at a school student audience. They have a great balance of text and pictures. Or go to the Life in Britain (2005) site which has posters on health, poverty, housing and employment. These are dramatic and visual, inspiring you to read the associated report.

Good posters explain a story with pictures. The fastest way to lose marks is to print an essay at 14 pt font and stick it onto card; expect instant failure no matter how good the academic content. Sound bite-length messages are wanted. However, this is still an academic exercise, so a soundbite alone will not do. Good academic argument and evidence, references and sources, are required.

Jolly-shaped posters, a volcano, map of Iceland, greenhouse for global warming or a county for planning policy will attract attention. Do not let the background overwhelm the message; the background should complement and enhance, not dominate. It is important that the background and illustrations do not bias your message with inappropriate, clichéd images (Vujakovic 1995). Amongst a gathering of 50 posters on 'Sustainability', there might be 45 on green boards with green, cream and brown mounts. Using a different background might make your poster stand out. Whatever the shape, ensure the maximum width and height are within the maximum size guidelines.

Interactive posters with pop-up effects, wheels to spin, fish that move as they migrate upstream, boreholes ejecting gases, or nests of overlays that slide away can be very effective. Take care that the structure can cope with handling. Use strong glue, reinforce cut edges and ensure the poster can be safely attached to the wall or board. Multi-layered card is heavy and likely to fall off.

> **SPELL CHECK EVERYTHING**

Highlight pictures and text with contrasting colours and backgrounds. Theme the colours, primary information on one colour, supplementary information on another. Be consistent in design format; it assists the reader. Let the story flow, for example, by placing argument or background information to the left of an image, graph or picture, and the interpretation or result to the right. It may be effective to have a hierarchy of information with the main story in the largest type, and more detailed information in smaller types. Colour- or size-coded information consistently helps the reader to decide whether to read the main points for a general overview, or the whole in detail. Edit hard to give main messages and change layouts around. A design styled like Figure 26:1 will take about 300 words, roughly one side of typed A4. The main story is in your report.

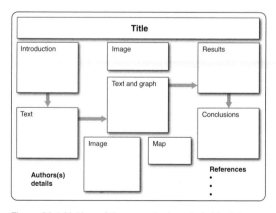

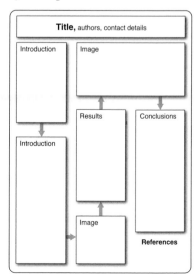

Figure 26:1 Neither of these poster layouts is ideal, how would you improve them? Each design suits about 300 words

There are costs involved in paper poster production, including card, photocopying, enlarging and printing photographic images. Colour printing is expensive. Be certain everyone is happy with the size and shape of each diagram before printing. A rough draft or mock-up, in black and white, before a final colour print, saves money. The first five to seven drafts are never right; font sizes that seem enormous on a computer screen look small on a poster board viewed from two metres away. Any poster produced by a group will be the subject of much discussion and change before everyone is happy! (See Figure 26:2.)

	Activity	Tasks and Issues	Dates
1	Focus and title	Get this clear. Draft a title, agree deadline dates. **Read the assessment rules in the module handbook.** Poster size and shape, landscape or portrait, and font sizes are often decided for you.	
2	Content	Get the academic content right. Research – discuss – research – discuss – spot the gaps – research – and then decide on focus and title.	
3	Draft plan A	Mock up a first layout. Does the title 'grab' the reader? Does it make the right academic points clearly? Has the design got simplicity and impact? Can it be read from 2m?	
4	Draft plan B	Show the mock-up (draft) poster to people. Ask what message it conveys. Is it what you intended? Are they following through the material in the order you wish? Can they distinguish major points from support material? What do they feel about the colour scheme? Are the font sizes large enough? Does the overall effect encourage them to read further? Are the main arguments supported by clear evidence?	
6	Final version	**Recheck the guidelines.** Double-check that all sources are acknowledged. Add keys, scales and titles to all maps, graphs, diagrams and photographs. Last chance to change the content and design to meet the criteria for a higher grade. Can you read it from 2m? Self-assess the final version (Figure 26:3). **SPELL CHECK EVERYTHING AGAIN.** Finalise, **Print,** Submit.	

Figure 26:2 Key steps in producing a group poster

E-posters are created in Word, PowerPoint and any graphics package. They are low cost, easy to store and shared electronically. They can be created by groups when people cannot meet. As with paper posters, consider the size on the screen. The maximum size will probably be defined as part of the project. Use attractive and clear typefaces and font sizes. Think about whether you find capitals easier to read or a mixed case style. Different typefaces, colours and shading can highlight different types of information.

An e-poster can include active HTML links, video, animated cartoons, and short interactive computer programs to demonstrate a model in action. There are many possibilities. The facilities and the expense of scanning material into digital form will govern your creativeness. Beware limits on file size, which may be set as part of the exercise. Be sure that an assessor can see your content; don't provide something that only runs with specialist software on your own computer.

Check and recheck assignment rules. See Figure 26:3 for assessment thoughts.

Group Names...

Poster Title...

	1	2:1	2:2	3	Fail	
Poster Structure (20%)						
Well organized						Disorganized
Focused						No clear focus
Research and Argument (60%)						
Main points included						No grasp of the main issues
Points supported by evidence						No corroborating evidence
Good use of relevant examples						Lack of examples
Ideas clearly expressed						Muddled presentation
Design and Presentation (20%)						
Creative layout and design						Poor design
Good graphics						No/poor graphics
Good word-processing						No/poor word-processing
Key points readable at 2m						Key points unreadable at 2m
Feedback						

Figure 26:3 Self-assess your poster

26.2 Instant posters

Some lecturers will ask you to create posters on flip chart paper or one PowerPoint slide in anything from 20 minutes to six hours. This is great practice for pulling ideas together quickly. Aim to express your ideas in as few words as possible; let pictures, diagrams and graphs make your points. Creating a poster to summarise the research outcomes from a fieldwork day happens on many GEES field classes. In geology your poster might include field sketches, data, graphs and, crucially, your interpretation and conclusions from your results. Most field days produce too much evidence. You cannot include every interviewee's feedback. Edit, edit, edit. Choose one quote that makes your point, the rest go into your report. Embedding photographs may be a great idea; make sure they are properly titled and that their relevance to your research is clear.

Looking at the posters created by other groups will help you to develop further ideas. Peer review and assessment of posters is usual because lecturers know that students learn much more from seeing how other groups express their ideas than from written feedback on their own poster.

26.3 Organizing an exhibition stand

Stands are frequently used for promotional purposes on open days and by university societies. Creating an exhibition stand may be an assessment. This takes time and effort. It is best done as a team to get the benefit of a diversity of ideas and energy. You need a team to talk to many visitors

> It is also a skill to talk to people.

on the exhibition day. A team of five or six will have many ideas, all of which are reasonable and possible. Leave time for discussion so that everyone gets involved and contributes fully.

Take a critical look at an exhibition. Careers events on campus are great opportunities. Employers promote their organization with clear and attractive messages. Consider the materials used, how they are presented, and decide what you like and dislike, what works and does not work for you. Consider the size of photographs, font size on posters and literature, multicoloured backgrounds and the height of materials. What enticed you to certain stands? What encouraged you to talk with exhibitors? Was it the free tea or toffees? What can you adapt for your project?

Pictures and demonstrations entice visitors to the stand when all of the team members are talking with other visitors. Ensure there is a logical arrangement to your materials so that a visitor can follow through the exhibits in the order you wish. Place posters and the pictures to attract passers-by at eye level. If people will sit down, consider what they can see from that level.

> *It was easier to talk to one person than a whole room, as you can adjust to their reactions and feel they are listening.*

An exhibition involves verbal discussion of research findings with visitors. These conversations require you to adapt your promotional 'blurb' to match visitors' interests and attitudes. This skill is different from an oral presentation for a tutorial or seminar; it develops your ability to think and adapt the message as you speak.

	Activity	Tasks	Dates
1	Focus	Get this clear.	
	Task	Draft a title, sort the timescale.	
	Time	**Read the assessment rules in the module handbook.**	
2	Audience	Tutors, students, general public, employers, school students, and who else ... ? What is the audience background knowledge? Does the display need to work for one or many audiences?	
3	Facilities and space	What are the limitations? – ask about the size of tables and poster boards. Is there a power point for a computer or lights? Is there space for a game or quiz? Chairs will encourage visitors to sit down and discuss issues.	
4	Draft plan A	Collect evidence, pictures, graph ... Be creative.	
	Decide on a theme	Co-ordinate colours and fonts throughout. Create draft version of display.	
		Re-check assessment rules.	
5	Draft plan B	Critically review plan A, collect further evidence, data, pictures to fill the gaps and revise the plan	
	Conversations	Practise your presentation/conversations to make sure you sell your ideas effectively	
6	Final plan C	Critically review plan B and conversations. Finalize posters, stand materials, and interactive elements. Practise conversations again.	

Figure 26:4 Key steps in working in a group to create an exhibition stand

It is quite normal in GEES schools for potential employers to be invited to exhibition-style events. Expect to see Environment Agency, local government officers or consultant engineers in the audience. Imagine that your display will be viewed by your first employer; aim to make a brilliant impression (Figure 26:4).

If this is part of an assessment, check the assessment guidelines or Figure 26:5.

Getting your 'storyline' right takes time. Practising before an 'assessor' turns up is a good idea. Listening is important, watch for body and language clues and tailor your message to your visitors' interests and experience. The following reflective quotes are from Level 2 geographers, after completing a two-hour exhibition for non-geographers.

> The most important thing the stand exercise helped me to learn is ... *'Whilst talking to people new thoughts about skills and proof for them came to you'. 'It is a skill to talk to people and to have to think on the spot, you have to adapt to different people and questions, you cannot just spurt out the same material'. 'I now know that I need to think through what I want to say in advance, so the message is really clear'. 'I learned a lot about interview technique, so that when people fire difficult questions at me I can deflect them and try to answer them without looking visibly flustered – therefore very helpful'.*

> I most enjoyed ... *'Creating the actual stand'. 'Getting lots of ideas from the other team members'. 'Seeing how good other people's presentations were'. 'Feeling a sense of achievement when someone enjoyed looking at the stand'.*

> I least enjoyed ... *'Standing around waiting for people to talk to us, feeling like a spare part'. 'Being asked quite difficult questions because it was difficult to think on your feet and there was only limited time for research'. 'Some of the people were really difficult to talk to'. 'Talking to the first person, but it got better as time went on'.*

Organizing and running an exhibition stand is great fun and very exhausting. Keep the chocolate handy!

Stand Title...	
Group Names..	
Please grade on a 1-5 scale, where 1 is Useless, 3 is Average, 5 is Brilliant.	
1. How well did the group articulate the principle points and issues? Comments:	1 2 3 4 5
2. How broad and effective was the evidence supplied to justify the claims? Comments:	1 2 3 4 5
3. How well did individuals use personal illustration and anecdote to support the claims? Comments:	1 2 3 4 5
4. How well did the stand materials support and enhance the points being made? Comments:	1 2 3 4 5
5. How creative were the ideas? How much impact did they have? Which materials or approaches would you commend? Comments	1 2 3 4 5
Additional feedback:	

Figure 26:5 Assessment criteria for an exhibition stand

26.4 References and further reading

Illinois University 2009 How to build a great poster, www.library.illinois.edu/rex/guides/ethnography/poster.html   Accessed 17 February 2011

Life in Britain 2005 Life in Britain, University of Sheffield, www.sasi.group.shef.ac.uk/research/life_in_britain.htm   Accessed 17 February 2011

Macaulay Institute 2010 Soils and Vegetation posters, www.macaulay.ac.uk/soil-posters/index.php   Accessed 17 February 2011

Macrae, E, Bond, C and Shipton, Z 2010 Elicitation of a Geological Model: based on intuition or solid principles? www.gees.ac.uk/events/2010/geobarriers10/documents/EuanMacraePoster.pdf   Accessed 17 February 2011

Mandoli, DF 2007 How to make a great poster, www.aspb.org/education/poster.cfm   Accessed 17 February 2011

Newcastle University 2009 Poster Presentation of Research Work, http//lorien.ncl.ac.uk/ming/dept/tips/present/posters.htm   Accessed 17 February 2011

Purrington C 2010 Advice on designing scientific posters, www.swarthmore.edu/NatSci/cpurrin1/posteradvice.htm   Accessed 17 February 2011

Vujakovic, P 1995 Making Posters, *Journal of Geography in Higher Education*, Directions, 19, 2, 251–256

Drop out 3

Remove one letter from each column to make one word, and the remaining letters align to make six additional words across. Answers on p 305.

S	U	R	P	H	U	R
E	T	L	A	N	R	L
M	E	H	S	I	O	Y
D	A	N	I	U	T	L
C	E	P	N	T	A	S
P	H	E	R	O	L	Y
E	N	T	C	O	P	Y

Jobs and careers for GEES graduates

I think I want a career, but maybe what I want is pay cheques.

Higher education changes people – attitudes and values develop and perspectives evolve. OK, I agree that getting a job is not a study skill, but having great skills will get you a job. What you need is the skills to explain the skills you have to the employer who wants your skills. Knowing what you want to do when you graduate is difficult and there is an enormous choice available. Your first advice point is your careers centre, or a web search for careers + service + university. Make time to think about what you enjoy, find rewarding, what you dislike and ask: 'How does what I know about myself, my skills and my attitudes, affect what I want to do when I graduate?'

This chapter mostly references UK sites for researching career options – I'm sorry there is no space to reference all the parallel international organizations. They exist, check them out. The Country Profiles on the Propects (2011) website is a great starting point for information about employment in countries from A to Z (actually Australia to the USA).

27.1 Jobs, employment and/or a career: start now

> *I went to Uni, had a lovely time, and am back home with mum and dad. It's OK but makes your degree seem a bit worthless ...*
> *(Graduate 2008)*

So you want a job? It is a good life move and comes to most people eventually. Ignoring employment advice and research at university risks you returning home to continue with vacation work (at vacation pay) rather than kick-starting the rest of your life with new opportunities. Researching careers reveals exciting summer vacation and gap-year work opportunities.

> *I didn't go to the Careers Centre because that would have meant making decisions, but being home, doing my summer job full time is a bit rubbish really.*
> *(Graduate 2008)*

You know that all GEES graduates have brilliant work-related skills, BUT most employers are ignorant of the skills your degree has given you, and don't know you either. Remember that your degree makes you *very* employable. Most geography and environmental science graduates, and lots of geologists are making the most of their skills in careers that are not degree-related. Pursuing certain careers requires further training (management consultancy, IT developer, solicitor, nursing, catering), but it is all good fun, interesting and usually you are paid while you train. Look at the literature in your campus careers shop, on the web, and don't rule anything out at the start. Having selected some options, research them properly. Fully prepared, you will stun employers with a well-written CV (*curriculum vitae*) and in interviews.

 TOP TIPS

→ Start career thinking NOW to help with getting vacation and long-term employment.

→ Set aside some 'Career Thinking Time' each fortnight. Half an hour or so on the bus, in the bath ...

→ Ask people about what they do, and consider whether a similar role would suit you.

→ Drop into your Careers Centre every couple of weeks and see what is happening. Keep track of career possibilities in the back of your work file. Make notes as you go. Go to the employer events.

→ Create and update your CV.

→ Consider Figure 27:1 – what is important to you? What do you want?

Sorting out the opportunities

Very few people stay with one organization for life. People change jobs and organizations, in some cases every three to five years. Try to think around your 5-year, 10-year and 15-year plans. What do you need to do now to make something else possible in the future? For example, someone wanting to work in the charity sector on water supply problems might do an engineering hydrology MSc and work with a firm of consulting engineers before moving to work with Oxfam. Someone wanting to start their own business might plan to work in one or more large companies to get management skills and experience of different aspects of business, and to complete a part-time MBA first. Potential professional geophysicists will need a master's degree to get their first job. This type of planning helps you acquire the jobs and training that will give you practical experience and networks of contacts to support you in the future.

Figure 27:1 Career thoughts (created in Wordle 2011)

- What are your aims and ambitions?
- Do you want to be in a big organization with structured training, options to change direction and plenty of colleagues?
- Do you want the security of a well-known organization?
- Do you want to work for yourself?
- Does being part of a small- or medium-sized enterprise (SME) appeal?
- Have you already seen your ideal job?
- What sort of lifestyle would you like?

These are difficult questions to answer; create time to think. Your search for a career is just the first step. Some people opt for 'portfolio careers', mixing a series of remunerated and unremunerated appointments and fitting them around other opportunities, including family and leisure activities (Hopson and Ledger 2011, University of Kent Careers Advisory Service 2011). Have a go at **Try This 27.1**.

TRY THIS 27.1 – Assessing your skills and experiences

Ask your friends and family about the careers they think you will enjoy, and to describe the skills that they know you have. Think about activities and experiences that motivate you. What work experience should you seek? What makes you stand out? Make some notes about your ambitions and your skills. Keep these notes safe to read before interviews.

27.2 Research your career options

Knowing what you want to do when you graduate is difficult and there is an enormous choice of careers. Tieger and Barron-Tieger (2001) *Do What You Are* can be a good start.

Environmental scientists and everyone interested in environmentally minded opportunities should start with YouTube and the Work2savetheworld (2011) video clips; watch Carolyn Roberts explaining how to be a water scientist and work with the police to track dead bodies in rivers; Huw Griffiths describing his work as a marine biologist in the Antarctic; and Rachel Conti, an air quality officer in London – just three of a selection of very useful insights from people doing their jobs.

Geologists should start at the British Geological Society of London (2011) Careers in the Earth Sciences website and work through the links. There are links to studying earth sciences and voluntary work too.

Geographers should start with the Royal Geographical Society (2011) Support for Students website. You may have used Going Places with Geography at school, look at it again to prompt thinking about how your ideas have evolved. Follow up the Careers with Geography section, YouTube and the Work2savetheworld (2011) video clips.

Fancy teaching in the infant, junior or senior school sector? There are jobs available in the UK in private or public schools, and you can enjoy running field trips to the locations of your choice. UK students should see the Graduate Teacher Training Registry (GTTR 2011) website for details on courses and how to apply.

For an MA or MSc in geographical and related subjects, check out Rogers and Viles (2003) Chapters 52–60, which describe postgraduate opportunities around the world. Almost all masters degrees can be taken in one year or done part-time. Look at degree options in planning, civil engineering, information science, computing, management, sociology, politics, international studies ... ; or in applied areas: conservation, water management, applied meteorology and climatology, urbanism, tourism, regional science, environmental planning, society and space ... If you fancy three years' in-depth study of one topic, then research PhD opportunities. Ask the postgraduate officer in your department for information and advice.

Careers research

Let your PC take the load. Get organized. Use **Try This 27.2** and **27.3** to evaluate your options.

TRY THIS 27.2 – Researching careers

Being systematic in researching opportunities helps. Whether in your careers centre or surfing from your PC:

First pick a career or occupation you think you might enjoy, perhaps using Prospects Planner (2011). Then:

1 Make a list of the things you imagine that the job will involve.

2 Find job descriptions and literature and/or websites for three companies/organizations – what skills do they need? (Start at the occupational profiles on Prospects (2011) which contains graduate case studies.)

3 Make a list of the skills and the evidence you can quote to show you have these skills.

4 How did your original ideas about this occupation match your first ideas? Are there surprises?

5 Is this the right type of company for you? How do your skills and aspirations fit the package? Do you want to do the same job but in a larger/smaller company?

6 Now research another career. Back to point 1.

TRY THIS 27.3 – Unlikely options

It really helps to confirm your job or career choice by knowing what you are avoiding. Pick two employment opportunities that you feel are not right for you and check out the websites, start with Prospects (2011). Many people are surprised when they see what different organizations and careers involve and realize that these are potential options. Equally, you may not like what you find and confirm that this is not a career for you.

27.3 Making applications

Perfecting your CV and covering letter

You need a succinct, accurate, clearly laid out CV and a covering letter. The CV informs employers about your past and current qualifications and experience. The covering letter links to your CV; use it to say more about the skills and experience that make you the ideal person for the job. ALWAYS revise your CV and letter for every application, so you totally match the advertised requirements. Develop a source CV (Figure 27:1) which is a warehouse containing all of your material in all possible forms, it makes it easier to copy and paste together a targeted CV tailored to attract each individual employer.

The employer reading your CV does not know about you or your skills. Get these up front, the personal, transferable and academic skills. Look back to Figure 1:2 (see p 4) and check off your skills. You have given 20+ presentations to groups of 5–50 people. You have used slides and electronic display material. You are comfortable and skilled with different spreadsheet and word-processing packages, GIS and mapping packages, modelling, statistics, and hands-on laboratory skills. You have taught yourself to use software. Tell employers GEES graduates have these skills. Updating your CV throughout your degree course saves time.

Most advertisements list the skill requirements of the job; make sure that your CV and letter addresses each area. Use the essential and desirable skills as subheadings. Tailoring your CV to each job is vital (see Figure 27:1, **Try This 27.4 and 27.5**). For postgraduate and research posts, focus on your geographical, earth or environmental sciences skills; for generic jobs focus on transferable skills and the work and student experiences that are particularly relevant. Most importantly, make sure your CV is:

✓ two sides of A4 only;
✓ professionally presented – show off your word-processing skills;
✓ punchy and precise – no waffle please;
✓ consistent in explaining what you have done – don't leave time gaps;
✓ consistently ordered within sections – put the most recent activities and experiences first;
✓ Be very careful to *treble-check* every entry before sending online applications.

Personal Details	Full name, address, email, telephone number (UK Equal Opportunities legislation means date of birth, nationality and marital status are optional).
Education	University name, programme details, degree result and, if relevant, brief mention of projects/modules undertaken. Senior school name and results from final exams. For earlier years summarize briefly.
Work Experience	Work placements, vacation jobs, voluntary work, any permanent or part-time work. Place these in reverse order, the most recent first. For the most recent and relevant jobs, use two or three sentences to state the content or skills you used and developed. Be concise but relevant. For example if you're applying for a planning job, then your CV might contain a section headed 'Planning Experience' which showcases paid and voluntary planning experience that you acquired during your degree, your work or your sports and hobbies.
IT Skills	Highlight this area. Remember the packages used in practicals and projects, statistics, mapping, GIS, programming, Excel, Access, Word, PowerPoint, internet searching, electronic library skills ... You may not feel very expert, but many applicants have fewer skills. Explaining you have taught yourself to use software (such as PhotoShop) is valuable information for an employer.

Other Skills	Include driving licence, first aid, and language skills. Explain your transferable skills using examples rather than lists: '... showed that I can be persuasive and present my case well'. 'Being left to manage the ... on my own was a challenge that was daunting at first. But I was able to cope with all the demands that were made and reorganized my work to meet the changing needs of the clients'. 'I enjoyed solving problems, negotiating and aiming to ensure the outcome was satisfactory for everyone'.
Interests and Activities	Be honest and don't just make a list. 'I play tennis for the university, rugby for the second XV hall team and have enjoyed coaching juniors at my local school'. 'I play bridge for a local club, I co-ordinate transport to matches'. 'At present I prefer to concentrate on aerobic training for skiing but intend to play more squash in future'.
Referees	Cite two, one academic and one from work. Always ask permission. Giving referees a copy of your application helps them to write a more focussed letter on your behalf.

Figure 27:1 Guidelines for contents of a CV

Covering letters should be short, positive and relevant. State which job you are applying for, why you are interested and the skills and experience that you can offer. Show that you are a career-minded person by stating how you see your career developing. Expand on a couple of points from your CV that are highly relevant for the job: 'As you see from the CV, I have considerable experience of...' Don't waffle or repeat chunks of the CV. Get to the point and keep it short.

 TOP TIPS

→ Revise your CV and letter for every application to completely match the requirements.

→ Always write job applications in formal English. This may seem obvious, but as more applications are submitted electronically, it is easy to drop into a casual, email writing style, which will not make a good impression.

→ Draft your CV and letter, then pretend to be the human resource manager reading it. Does it imply an enthusiastic, energetic and creative person with oodles of skills? Redraft until it does.

Think of yourself as a marketable product. You possess many skills that employers are seeking; it is a matter of articulating them clearly to maximize your assets.

TRY THIS 27.4 – Action words for applications and CVs

Draft your CV and letter, then consider whether you are really expressing your skills fully. Try CV bingo – how many terms can you SENSIBLY include?

Achieve	Catalogue	Delegate	Forecast	Market	Produce	Retrieve
Accomplish	Chair	Demonstrate	Guide	Mediate	Promote	Review
Act	Clarify	Design	Generate	Meet deadlines	Propose	Revise
Adapt	Classify	Determine	Handle	Moderate	Provide	Schedule
Allocate	Collaborate	Develop	Help	Modify	Publicize	Screen
Advertise	Collate	Devise	Identify	Monitor	Publish	Select
Analyse	Collect	Direct	Implement	Motivate	Purchase	Serve
Advise	Communicate	Distribute	Improve	Negotiate	Qualify	Set up
Apply	Compare	Draft	Increase	Obtain	Raise	Solve
Administer	Compile	Edit	Influence	Operate	Recommend	Sort
Appraise	Complete	Employ	Initiate	Order	Record	Staff
Approach	Compute	Encourage	Innovate	Organize	Recruit	Summarize
Approve	Conduct	Ensure	Install	Oversee	Redesign	Supervise
Arrange	Configure	Establish	Instruct	Participate	Reduce	Support
Assemble	Consult	Estimate	Integrate	Perceive	Regulate	Survey
Assess	Contract	Evaluate	Interpret	Perform	Renew	Synthesize
Assign	Control	Examine	Interview	Persuade	Reorganize	Teach
Assist	Co-operate	Execute	Introduce	Plan	Repair	Train
Attain	Co-ordinate	Expedite	Invent	Prepare	Report	Transmit
Audit	Correct	Facilitate	Investigate	Present	Represent	Update
Budget	Create	Familiarize	Lead	Prioritize	Research	Validate
Calculate	Decide	Formulate	Manage	Process	Resolve	Work with

Use these sites for guidance and examples of CVs. All sites accessed 15 February 2011

Association of American Geographers (follow the Jobs & Careers link) http://www.aag.org/

Careers in GIS http://www.gis.com/content/careers-gis

Earth Science Jobs http://www.earthsciencejobs.co.uk/jobsite/

Earthworks jobs http://www.earthworks-jobs.com/

Environmental science: Your skills, Prospects, http://www.prospects.ac.uk/options_environmental_science_your_skills.htm

Geography: Your skills, Prospects, http://www.prospects.ac.uk/options_geography_your_skills.htm

Geology: Your skills, Prospects, http://www.prospects.ac.uk/options_geology_your_skills.htm

How to write a great Curriculum Vitae, The University of Waikato, http://www.waikato.ac.nz/sasd/careers/articles/cv.shtml

Job applications: Selling your skills, Prospects, http://ww2.prospects.ac.uk/cms/ShowPage/Home_page/Applications_CVs_and_interviews/Job_applications/Selling_your_skills/p!eXfdpk

Making Applications, Newcastle University, http://www.ncl.ac.uk/careers/jobs/applications/cv.php

University of Washington Department of Geography, Career Resources,http://depts.washington.edu/geogjobs/

The interview

The challenge is to prepare properly, otherwise your effort with application forms and travelling is wasted. Look at video clips of interviews for advice on what to do and not to do. YouTube has some great examples. Read and reread company literature; read everything they send; read and reread the advertisement; map your skills to their requirements. Have more examples ready. Most questions are straightforward; the trick is to have thought about the answers. Start preparing answers to:

- Why are you choosing to be a ... ?
- What can you offer to ... ? (human resources, marketing, software sales)
- What prompted you to apply to this organization/sector/company/country?
- What do you know about this business/sector?
- What interests you most about this job?
- What particular skills do you have which will help you make a success of this job?

- Why do you want to work in a large/middling/small organization?
- Why should we offer you a job with us?
- Where do you see yourself in three years'/five years' time?
- What were the three most significant/tough/traumatic things that happened in your year abroad/life? How did you manage them?
- Variations on: What are your strengths and weaknesses? What can you offer the company?

An interviewer for most commercial jobs has no interest in your knowledge of coasts, boreholes or rural development. But he might want to know that 'the experience of generating a group report on population decline with four people helped me to understand the importance of planning, drafting and agreed editing, and integrating different people's perspectives. It was tough because of the deadline, which we met, but it meant that next time we worked like that we were better at dividing the tasks and criticizing and redrafting'.

Use real examples all the time: 'In a group project (mapping/fieldwork ...) I organized ... , negotiated for ... , shared the decision-making ...' 'Mentoring secondary-school students through the campus programme to create their own webpages was really rewarding. I realized I liked working both one-to-one and with small groups and that boosted my confidence to ...' 'I was elected treasurer of the ski club, which meant persuading people to part with subs, negotiating for union support, creating ... , supporting the committee by ...' Many university careers services use the Association of Graduate Careers Advisory Services' (AGCAS) video guides to help students with interview technique and career advice. (Search online for 'AGCAS video guides to interview and assessment centres' , to find a link.)

Chasing a job may involve you in any or all of: telephone interviews; selection centres; employer interviews on campus; campus careers fairs; recruitment fairs; career workshops; employer presentations; psychometric tests; standard application forms; employer's application forms and ... Use the experts in your careers centre to make sure you make the most of opportunities. Are there practice interview sessions? Will careers centre staff check your application form and CV before you post it? Bookmark the local careers information web sites and check them regularly.

27.4 Getting placements/internships/being paid

Summer work that is paid at a good rate and gives you insights into the career you think you want, is exceedingly valuable, and can help significantly to boost your CV. The best advice is 'plan ahead'. These jobs are sought after and need applications up to a year in advance. You may find that you hate the company and never want to be a banker or curator or journalist, but your CV is boosted and you will not waste the first couple of years at work on something that doesn't suit you. In the best case, you may get job for next year. Placements don't necessarily pay, but some companies and schemes offer travel expenses and a few pay very well. UK students should check out the STEP scheme; it is competitive but they are very well organized, pay and have prizes. Check out **Try This 27.6** to get a feel for students' experiences and **Try This 27.7** to find volunteering opportunities.

⚙ TRY THIS **27.6 – Student placement**

Here are four short reports on students' experiences of placements. What are the benefits they are reporting? What skills could they add to their CV?

Hannah Goldstein: I was thinking maybe library work, maybe museums, but I really don't know what. I contacted my local area museum, having found them on www. museumsassociation.org/home and they were really helpful. I had to send them my CV and a covering letter, saying what type of museum I was particularly interested in – there are big differences between a big interactive museum and a small traditional display case one – and how much time I could give them. In the end I opted for a placement with my local county museum, which specializes in agricultural history. I offered four weeks during the summer. I wasn't paid, but got my travel costs and a meal allowance. I was taking phone calls from the first hour, working with visiting schools and disabled youth groups, helping with enquiries and staffing the information desk. I had a week shadowing the marketing manager, which meant I went to some interesting meetings as an observer. It really helped me to understand the business side. It was interesting work, and they asked me to come back, but it was difficult with me not earning anything while I was there.

Colette O'Connell: I had asked to work at our local newspaper and printing offices. I already had quite a lot of experience with my university newspaper on the features side, and I used to edit my school magazine, so I was expecting to contribute quite a bit. Human resources were fine about arrangements in advance, but when I turned up I was treated like a 14-year-old volunteer. I was ignored for most of the first morning. I got to do some photocopying and to help sort the post. I tried to involve myself in the news room, but they were very short-staffed and people didn't seem to have the time to help me. I explained that I had experience with newspapers, but the message didn't get through. Basically, no one knew what to do with me, and although I tried to talk to the human resources people, they were in meetings. I stopped going after three days.

Abdul Haq: I applied to the STEP programme after Easter through the careers service at my university, and I was matched to a placement in my home town. I had three interviews to get my placement and it was really great. You can find out all about it on www.step. org.uk. You get paid on STEP, I did ten weeks in the summer vac. They wanted a whole new section of their web pages sorted out, so I had to learn how to do that. I did the basic research to get the information I needed and to cross-check it for accuracy. I found out how to do layouts and access the corporate web design material so my new pages fitted with the company image. I had regular meetings with my supervisor, so we were both clear about the aims and my progress. She was very helpful, and checked everything before it was uploaded to the web. I was visited by somebody from the university to make sure the STEP placement was working out for me and the company. Besides gaining web design and writing skills, I gained lots of office skills, and used my research skills conducting telephone interviews and a web questionnaire to get my data. My project was selected to go forward for the region to the national STEP finals, so I won £500 too, as well as being paid. Great experience.

Jamie Alexander: I want to work in business, but I wasn't sure where so I applied to quite a lot of blue-chip organizations offering summer internships. It was very competitive. I had to apply online using the same form that they use for graduate recruitment, and then I had to take aptitude tests, and have an interview with a line manager. I was offered a summer internship, which was really very well paid. To a certain extent, an internship is like an extended interview: they were seeing if I would be a good employee and I wanted to know if they were right for me. I was given a specific project to do, which culminated in a report to the director of marketing, as well as going to meetings, drafting letters and dealing with clients directly. Over the summer they moved me through four departments, so I made lots of useful contacts. I realized working for an organization of this size has lots of dimensions, and that there is great scope for promotion to different parts of the business. I went on quite a few internal training courses, and people were helpful and took time to explain what was happening to me. I applied to them in the autumn when I was back at university, and I have been offered a job to start next summer.

> **TRY THIS 27.7 – Researching volunteering**
> What volunteering opportunities are arranged by your student union or careers centre?
> Fit something around your work in term time and add to your CV.
> Check out opportunities online using: internship, volunteering, placements.

27.5 References and websites

British Geological Society of London 2011 Careers in the Earth Sciences, http://www.bgs.ac.uk/vacancies/careers.htm   Accessed 15 February 2011

Cantwell M 2010 Portfolio careers for people in their 20s and 30s, Career Shifters, http://www.careershifters.org/expert-advice/portfolio-careers-for-people-in-their-20s-and-30s   Accessed 15 February 2011

GTTR 2011 Graduate Teacher Training Registry (GTTR) http://www.gttr.ac.uk/ Accessed 15 February 2011

Hopson B and Ledger K 2010 Portfolio Careers, http://portfoliocareers.net/ Accessed 15 February 2011

Prospects 2011 Country Profiles, www.prospects.ac.uk/country_profiles.htm Accessed 15 February 2011

Rogers A and Viles H (Eds.) 2002 *The Student's Companion to Geography*, 2nd Edn., Blackwell, Oxford

Royal Geographical Society 2011a Support for Students, http://www.rgs.org/OurWork/Research+and+Higher+Education/Support+for/Support+for+Students.htm   Accessed 15 February 2011

Royal Geographical Society 2011b Careers with geography, http://www.rgs.org/OurWork/Study+Geography/Careers/Careers+with+geography.htm Accessed 15 February 2011

The Institution of Environmental Sciences, 2011 Supporting Careers, http://www.ies-uk.org.uk/resources/careers.php Accessed 15 February 2011

Tieger PD and Barron-Tieger B 2001 *Do What You Are: Discovering Your Perfect Career*, 3rd Edn. Little, Brown & Company, Boston

University of Kent Careers Advisory Service 2011 I want to work in … a different way, http://www.kent.ac.uk/careers/alternatives.htm Accessed 15 February 2011

work2savetheworld 2011 YouTube clips, http://www.youtube.com/work2savetheworld#p/ Accessed 15 February 2011

Wordle 2011, http://www.wordle.net/ Accessed 20 November 2010

Volunteering opportunities for UK students

Ecovolunteer 2001 http://www.ecovolunteer.org/ Accessed 15 February 2011

Fledglings 2010 Fledglings graduate placements network, http://www.fledglings.net/ Accessed 15 February 2011

Step 2010 Step nurturing talent, http://www.step.org.uk/ Accessed 15 February 2011

Talent ladder 2010 Talent ladder: what you need when you need it, http://www.talentladder.com/ Accessed 15 February 2011

Volunteering England 2010 Volunteering England, http://www.volunteering.org.uk/ Accessed 15 February 2011

VSO 2010 Voluntary Service Overseas UK, http://www.vso.org.uk/ Accessed 15 February 2011

Words in Geo-words 5

Set 8 minutes on the timer (cooker, mobile …) and see how many words you can make from:

TECTONIC

Answers on p 305.

28

'Of shoes, and ships, and sealing wax ...' GEES student essentials

Here is a pot-pourri of items that are handy for GEES students. Some may seem bizarre and irrelevant, but check out the headings so that you can find them later.

28.1 Who is the president of Portugal?

Lecturers tend to assume students have working knowledge of current UK, Europe and world leaders. Who's who and who's where? Look at **Try This 28.1** and bring yourself up to speed if you don't know all the answers.

TRY THIS 28.1 – Who's who?

Who's who internationally?	Who are the UK's/Your country's leaders?
President of the United States of America	Prime Minister
Vice-President of the United States of America	Foreign Secretary
Is the US President a Democrat or Republican?	Home Secretary
Chancellor of Germany	Chancellor of the Exchequer
President of France	Leader of the Opposition
Prime Minister of France	Shadow Chancellor
Prime Minister of Japan	First Minister for Wales
President of Germany	First Minister for Scotland
President of China	Governor of the Bank of England
President of Israel	Archbishop of Canterbury
President of Russia	Which political party is in power?
Prime Minister of Russia	
Secretary-General of the UN	

28.2 Word-processing – do you know?

Figure 28:1 is a check list of WP skills. Take 30 minutes to explore what the package can do to help you work smarter. ✓ them off as you go, and celebrate a full list.

Insert page numbers	Vital in every document, please add.	✓
Spell and grammar checker	Set it to always be on at the highest level. The suggestions are not always what you want but will make you think about your writing style.	
Language	Make sure your documents are spell checked in the right language. English (UK) is not the same as English (US).	
Bulleted or numbered lists	Helps the reader to understand the message. Insert jolly ☘ ✓ ✗ ☺ symbols as in this textbook too, where appropriate.	
Images	Photos, diagrams, charts, clip art; cite the source correctly, pictures usually save hundreds of explanatory words.	
Header and footer information	Important in longer documents. Put your name, student number, module code and title, in the footer, with the page numbers.	
Tables	This book has many tables to make information clearer. Tables are not just for numbers.	
Titles and sub-headings	Use larger font sizes to clarify sections and structure your writing. Subheadings help you see if your answer is balanced: a three-part question leads to three subheadings of roughly equal weight. A few tutors dislike subheadings; write the essay with sub heads and delete them before submission.	
References	Helps to create the reference list, keeps track of entries and linked to EndNote or similar packages, it saves time when finishing documents.	
Footnotes	More often used by historical and cultural geographers than in the sciences, footnotes add more detailed explanations, but are not needed in tutorial essays.	
Thesaurus	Stuck for the right word? See what else is suggested.	
Review and track changes	These commands are essential for group writing, so people can see who has added and changed each section.	
Columns	Useful for newsletter, handbill and PR writing. A two-column table will produce a similar result.	
Sort	Essential for getting references into alphabetical order in every essay. And good for any other ordering tasks.	
Line spacing, fonts, justification, colours, borders, margins …	OK – you have known all this for at least ten years, but are you using it to give you the best design. Aim to have a professional-looking document every time.	

Figure 28:1 Checklist of word-processing skills

28.3 IT speak

Here is a starter list of useful terms. Add to it.

Backing up	Making copies of files on PC, USB stick and … so you do not lose your work when your computer fails.
Blog	An online diary – be careful with what you say if it can be read by anyone. If you would be embarrassed if your granny read it, don't post it.
Bookmark	Find your favourite web pages quickly.
Cookies	Software programs you won't notice that allow websites to track their users.
Domain	Part of the URL (uniform resource locator) website address.
Firewall	Security programs that can restrict links between computers and internet services.
FTP	File Transfer Protocol, lets you move large files across the internet.
HTML	HyperText Markup Language. The predominant markup language for web pages.
ISP	Internet Service Provider.
Open Source	Applications and programs that are freely available for you to use as you need.
Portal	A web page that gives access to further sites. Many universities have a portal giving student and staff access to the university intranet, and links to the VLE.
Spam	Junk mail that clogs your email inbox.
URL	Uniform resource locator, i.e. website address.
USB	Universal serial bus. The computer port that accepts high-speed connection devices such as USB sticks, external hard drives and jolly office USB toys and gadgets.
VLE	Virtual Learning Environment. An online space for module and course information, learning materials, support materials, blog and discussion sites, and many other exciting activities.
WiFi	Process for connecting to the internet through wireless technology.
Wiki	A discussion site where many people can add their thoughts, opinions and ideas.
Zip	A compression process that reduces file sizes for efficient storage and transfer.

28.4 Dictionaries and atlases: do I need one?

It is the consensus of a non-representative, random sample of students who walked into my office that you need *a general dictionary* and *a thesaurus*, or know how to access online versions. Buying a geography, earth sciences or environmental sciences dictionary is less vital; they are in the library. The advantage of subject dictionaries is that they contain short explanations on topics, making them much more informative than a general dictionary. They can be a good starting point for research. A general dictionary does help.

Atlases are useful but a good one is expensive – the university library has plenty and Google maps are available. An atlas is not a vital purchase, but if you are doing regional geography options, you must make the time to check out and learn the names of relevant cities, rivers and regions. For modules on Japan, China and the CIS, this is a non-trivial activity. What is vital is to check the edition date: there have been many place name changes in the last 30 years. Many online maps are out of date for eastern Europe, for example.

If your gran wants to give you something useful for Christmas, then an atlas is a good wheeze and it helps with holiday planning.

Dictionaries available in a library near you include:

Allaby M 2008 *A Dictionary of Earth Sciences*, 3rd Edn, Oxford University Press.
Gregory D Johnston RJ, Pratt G, Watts M and Whatmore S (eds) 2009 *The Dictionary of Human Geography*, 5th Edn., Wiley-Blackwell.
Mayhew S 2009 *A Dictionary of Geography*, 4th Edn., Oxford University Press
Porteous A 2008 *Dictionary of Environmental Science and Technology*, 4th Edn., Wiley

Always use your spell checker but 'spill chequers due knot awl weighs git it write'.

28.5 Accuracy, precision, decimal places, uncertainty, bias and error!

Most GEES data are noisy, imprecise, inconsistent and may be biased. The trick is to recognize sources of error, acknowledge them and discuss their impact, for example, in the methodology or discussion sections of a report. Discussing error is not a matter of mega *mea culpa*, 'it was his/her fault' statements. Unless exceptional care is taken, items, questions and variables are forgotten, data collection methods influence the measurements and instruments may be inaccurate. The idea

> **Accurate value:** a measurement that is as close to the right value as possible.
>
> **Precise data:** can be reproduced consistently but are not necessarily accurate.

is to minimize all possible errors, but only the infallible will succeed. Aim to be as objective as possible.

A true value, the absolutely accurate, correct value of something can be hard to determine; more often one has the best estimated value and one can describe it in terms of its accuracy and precision.

A grid reference position on the ground can be located with increasing accuracy as the scale of the map increases. On a 1:50,000 map, a position may be in error by ±32m, whereas from a 1:25,000 map, the error may be ±15m. The accuracy of points located from satellite imagery will depend on the height and type of the platform and the spatial resolution of the images, and biased by topographic distortions and off-nadir angles during scanning. The accuracy of a global positioning system (GPS) will depend on the sophistication of the GPS instrument and the number of satellites within view.

It is good practice to quote a value with its associated resolution, for example, with error bars and scatter plots, giving the reader a feel for the precision and accuracy of the information. A location may be given as X ±16m. An average, 72.25 per cent, may be reported with the range 66.3 per cent to 78.2 per cent. Temperature data from a thermometer, which can, at best, be read to 0.2 of a degree, would have the uncertainty associated with a reading expressed as 8.2 ±0.2°C. Measuring temperature with a thermocouple should give greater accuracy, because the measurement can be made to the nearest 0.01°C. It is therefore a more precise measurement, 8.23 ±0.01°C.

However, if the thermocouple consistently under-reads the thermometer, although the data are more precise, they would be less accurate than those from the thermometer. The readings would be precise but inaccurate, due to instrumental error. Similarly, if a GPS is not properly calibrated, subsequent readouts will be inaccurate.

Decimal places

A disadvantage of digital technology is that calculations can be reported to many decimal places. You must decide what is relevant for your study or instrument. Avoid the absurdity of 'a family with 2.45678 children', or, 'Therefore we conclude the residents of Hobbiton make 3.21776 shopping trips per week to Buckland', or 'Survey results show the population of Stats-on-Sea are 26.3732 per cent Hindu and 56.4567 per cent Christian'.

How many decimal places to quote? Think about the accuracy you require and the application. An average temperature of 16.734567°C is experimentally accurate and appropriate for a physicist doing neutron experiments; for a soil temperature, 16°C or 16.7°C will do. A digital pH meter used in the laboratory or field may show two decimal places, but the digits are unlikely to settle because pH is not a stable measure. In practice, a soil pH will have local variability and to report to better than 0.2pH has no practical pedological value.

Is the sample biased?

> **Bias:** tendency to express one view or opinion without presenting alternative views.
>
> **Statistical bias:** occurs in a number of ways – especially when data is selected such that not all samples could be equally selected, or a hypothesis is tested which is not independent.

Bias is a consistent error is in data. There is plenty of information in statistics, laboratory and fieldwork sessions about taking samples in the right way and with enough replicates so that bias is minimized. Think about potential sources of bias when considering results. I think you would agree that sampling two midday temperatures in July at Filey in 2003, and using these data to describe the average climate of the UK in the twentieth century would leave the reader unimpressed. Such data are obviously inadequate and biased. Does your data set have similar, albeit less blatant, drawbacks?

Data entry errors

These are very, very, very, very common. Check your data very carefully every time against original field or laboratory notes. Using the spreadsheet command to show the smallest and largest datum can help to find the commoner errors: forgetting a decimal or typing two items as a single entry.

Computational errors

Have all the formulae for calculations been entered correctly? Cross-check the equations and look at the answers to make sure they are in the right 'ball park'. A quick, back-of-the-envelope calculation using whole numbers can give a feel for the range of answers to expect. If answers fall outside the expected range, check all the calculation steps carefully. Keep units the same as far as possible, and make careful notes when the units change though each stage of a calculation.

28.6 Latin words and phrases

'amos, amas, where is that lass?'

There are many Latin phrases in normal, everyday use. Understanding some of them will get you through university and assist in solving crosswords for life. They are normally set in *italics*.

a priori	reasoning deductively, from cause to effect
ad hoc	for this unusual or exceptional case
ad hominem	to the man, used to describe the case where an argument is directed against a person rather than addressing the case itself
ad infinitum	to infinity
ad lib; ad libitum	at pleasure; (speak) off-the-cuff
CV	curriculum vitae, a short description of your life, suitable for employers
carpe diem	seize the day
de facto	in fact, or actually
e.g.; exempli gratia	for example
et al.; et alia	and other persons; appears in essays to refer to multiple authors of articles or books
et seq.; et sequens	that which follows
etc.; et cetera	and the rest; and never used in a good essay; as it implies lazy thinking
ex officio	by virtue of office
honoris causa	as an honour
i.e.; id est	that is
inter alia	among other things
ipso facto	thereby, or by the fact
mea culpa	it was my fault
NB; nota bene	note carefully, or take note
op. cit.; opere citato	in the work cited
post hoc	used to describe the type of argument where because x follows y, it is assumed y causes x
reductio ad absurdum	reduced to absurdity
[sic] sic	thus; often printed in brackets to indicate that the preceding word or phrase has been reproduced verbatim
sic transit gloria mundi	so passes earthly glory (not Gloria's been ill on Monday's bus)
tempus fugit	time flies
vi et armis	by force of arms
viva voce	by oral testimony, usually shortened to viva; meaning an oral examination

and finally the motto for many a good fieldclass:

ergo bibamus	therefore let us drink

28.7 Greek alphabet

I thought I understood that rivers stuff, but what is this
lamb deer he goes on about?

Scientists especially will find lecturers freely flinging *thetas* and *gammas* around in lectures. Duck. The highlighted items appear regularly. You can write faster using symbols, making it worth knowing these.

A	α	Alpha	I	ι	Iota	P	ρ	Rho
B	β	Beta	K	κ	Kappa	Σ	σ	Sigma
Γ	γ	Gamma	Λ	λ	Lambda	T	τ	Tau
Δ	δ	Delta	M	μ	Mu	Y	υ	Upsilon
E	ε	Epsilon	N	ν	Nu	Φ	φ	Phi
Z	ζ	Zeta	Ξ	ξ	Xi	X	ψ	Chi
H	η	Eta	O	o	Omicron	Ψ	υ	Psi
Θ	θ	Theta	Π	π	Pi	Ω	ω	Omega

Every water resource scientist knows that Eureka means
'the bath water's too hot'.

28.8 Mohs scale of mineral hardness

In 1812 the Mohs scale of mineral hardness was devised by Frederich Mohs, a German mineralogist. He selected ten minerals of increasing hardness which were commonly available. Mohs scale is not linear. It is relative allowing geologists to compare the hardness of other minerals.

Hardness	Mineral	Associations and uses
1	Talc	Talcum powder. Very soft, it can be scratched with a finger nail.
2	Gypsum	Plaster of Paris. Gypsum is formed when seawater evaporates from the earth's surface.
3	Calcite	Carbonate mineral, limestone and most shells contain calcite.
4	Fluorite	Fluorine in fluorite prevents tooth decay. Halite mineral.
5	Apatite	When you are hungry you have a big 'appetite'.

6	Orthoclase	Orthoclase is a feldspar, and in German, 'feld' means field. Most abundant earth crust mineral.
7	Quartz	SiO_2 second most abundant mineral.
8	Topaz	The November birthstone. Emerald and aquamarine are varieties of beryl with a hardness of 8.
9	Corundum	Sapphire and ruby are varieties of corundum. Twice as hard as topaz.
10	Diamond	Used in jewellery and cutting tools. Nice in rings. April birthstone.

28.9 Plant names

Where did houseplants live before urbanization? No houses, no houseplants.

Plants have Latin names because the local or common names for plants change as you move around a country. Botanists reclassify plants as more is known about their biology and genus, so an up-to-date flora is vital.

All Latin names are indicated by underlining in handwritten text and by *italics* in typescript. When a plant's genus is referred to generally, rather than a particular species, the convention *Carex sp.* is used. If there are lots of species of the same genus, then use *Carex spp.* Some plant names are long and it becomes tedious to repeat them every time in full; the convention is to use the full name initially, *Carex nigra*, and thereafter *C. nigra*. Be careful, this convention refers to the last mentioned plant.

Find a reference guide to flora that you like in a book shop and make sure it fits your jacket pocket for fieldwork. There are a range of useful plant and tree identification apps too. See the apps Store for the latest and take a look at some of these titles:

Grey-Wilson C 2000 *Wildflowers of Britain and Northwest Europe*, DK Handbooks, Dorling Kindersley Publishers Ltd, London

Fitter A and More D 2004 *Trees*, Collins, London

Fitter R, Fitter A and Blamey M 1996 *The Wild Flowers of Britain and North West Europe*, 5th Edn., Collins Pocket Guide, Harper Collins, London

Hubbard CE 1992 *Grasses*, Penguin, London

Phillips R and Grant S 1980 *Grasses, Ferns, Mosses and Lichens of Great Britain*, Pan Books, London

... and remember, Blewitt, Cramp Ball and Jelly Babies are fun guys to be with!

28.10 'We didn't do this at school'

Geography, earth and environmental sciences are all big and diverse subjects; no one gets to do all of it, at school or anywhere else. No problem. All three disciplines have traditionally recruited excerpts and ideas from other subjects.

 TOP TIPS

☺ Don't panic.

☺ Go to the lectures and get the general idea of the topic and material. Sort out the focus of the module. Read the materials on the VLE, plan your reading strategy.

☺ Pre-university: Try local libraries for senior-school texts to pick out the background before a course starts, BUT you must move on to university-level texts. School texts are starting, not ending, points.

☺ Take school notes to university; your soil notes may help someone who can help you via their economics notes. People with a maths background are very popular in statistics modules.

☺ If you have difficulty with essay writing, and especially if you have not written an essay in three years, find someone who did English, history or economics in their final school years and pick their brains about writing. There are plenty of people who have excellent English language skills and can explain how to use semi-colons.

☺ Ask people which reading they find accessible and comprehensible.

28.8 Equations and maths!

The squaw on the hippopotamus is equal to the sum of the squaws on the other two hides.

If the sight of an equation sends you into a state of total panic, don't worry! No one will ask you to solve Fermat's Last Theorem. You knew all about equations at some stage, it's just a case of letting it flood back. The real problem with university is that the education you were given from 5–16 gets to have some pay-off. Unfortunately, if some things, like mathematics and chemistry, went into a bit of dormant memory, it is time to reopen it.

A lecturer will be talking about the latest excitements in 'property and landlords', 'river gravels' or 'sustainability in the Antarctic' with delightful slides and then diverts to equations. DO NOT PANIC (well, only a little). Scientific notation is shorthand. You can do the easy ones: EITHER 'Water added to a catchment from precipitation could be accounted for by evaporation loss, evapo-transpiration loss, runoff in the river, and changes in the soil and groundwater stores' (28 words) OR

$P = E + ET + Q + \Delta S + \Delta G$ (not 28 words).

It is essential to define the elements of the equation, but having done it once, time and space is saved later by not having to write it all out again. Where appropriate, WHICH IS 99.9999% OF THE TIME, the units of measurement should be included. Darcy's Law for water flow through a soil or sediment sample:

$Q = KIA$ where Q = Discharge of water though the soil
              K = Hydraulic conductivity of the soil (a function of water content)
              I = Hydraulic gradient
              A = cross-sectional area of the soil sample

will score 60%, but is better expressed as:

$Q = KIA$ where Q = Discharge of water though the soil $cm^3.s^{-1}$
              K = Hydraulic conductivity of the soil (a function of water content) $cm.s^{-1}$
              I = Hydraulic gradient   dimensionless
              A = cross-sectional area of the soil sample cm^2

for 100%.

Life gets to be more fun when data comes in the form of tables and matrices. A table that lists all the students in your class from Sheila to n, and the local pubs, The Orienteer and Compass to n, can be referred to using matrix algebra. Yes, algebra. DO NOT PANIC. What seems to cause confusion is making a distinction between the code used to refer to positions in a matrix and the actual data values themselves. Basically, it works like map reading with grid references. If the data in the matrix refers to cash spent by individuals in the pub over a term, as here:

HOSTELRY		The Lost Orienteer	The Lorenz and Lowry	The Mattock and Spade	The Hazard and Impact	... n
		i_1	i_2	i_3	i_4	... i_n
Sheila	j_1	£25.50	£16.75	£0.00	£3.00	
Wayne	j_2	£45.70	£0.00	£16.80	£6.50	
Spike	j_3	£52.80	£12.80	£36.45	£22.40	
Liz	j_4	£28.67	£24.87	£16.85	£35.78	
Jenny	j_5	£45.23	£12.60	£21.76	£0.00	
..						
n	j_n					

Each of the pubs is coded as i_i ... $_n$ and each of the class as j_1 ... $_n$. The total number of pubs is $\Sigma i=1$-n and the total number in the class is $\Sigma j=1$-n. You can refer to a single cell; the shaded cell for Liz in The Lorenz and Lowry is i2j4. The total amount spent by Sheila is the sum of all the columns i=1-n for row j_1. So that's the sum of $\Sigma 25.50 + 16.75 + 3.00 = £45.25$, and Liz has spent £24.87 in The Lorenz and Lowry.

When you are faced with an equation, read it carefully and recall the rules (e.g. solve elements in brackets first). If you are feeling out of control, verbalize the instructions as in the multiple choice below. Which of these verbal instructions are right?

$a+15b-e$ a) Multiply b by 15, then add a, and subtract e.
  b) The sum of 15 times b and a, minus e.
  c) a plus 15, times b and subtract e.
$6(x^3+y^2)$ d) Take the square of y and add it to the cube of x and multiply the total by 6.
  e) Multiply the cube value of x by 6, and then add the squared value of y.
  f) Cube x, then square y, add the two values together and multiply by 6.
$2x^2-y^2$ g) Two times the square of x, minus y squared.
  h) Square x and y, subtract and multiply by 2.
  i) Square x and multiply by 2, and then subtract the squared value of y.

The answers are a, b, d, f, g and i.

Percentage problems?

$50\% = \tfrac{1}{2} = 0.5$ $\quad\quad$ $33.33\% = \tfrac{1}{3} = 0.333$ $\quad\quad$ $25\% = \tfrac{1}{4} = 0.25$

$20\% = \tfrac{1}{5} = 0.2$ $\quad\quad$ $5\% = \tfrac{1}{20} = 0.05$ $\quad\quad$ $1\% = \tfrac{1}{100} = 0.01$

For further (and much more useful) memory jogging see:

Van Der Molen F and Holmes H 1997 Maths Help, in Northedge A, Thomas J, Lane A and Peasgood A, *The Sciences Good Study Guide*, The Open University, Milton Keynes, 301–395.

28.12 Chemical notation

You play the symbols, and I'll bang the drum.

Soils or water chemistry lectures can come as a shock if you have not cuddled up to the periodic table recently. The tables below contain the main chemical elements that feature in geophysics, mineralogy, geomorphological, hydrological and soils modules. Use them as an *aide-mémoire*. They are simply convenient, internationally recognized abbreviations, and nothing more. Add others as you encounter them.

Symbol	Name	Symbol	Name	Symbol	Name
Al	Aluminium	Fe	Iron	P	Phosphorous
C	Carbon	H	Hydrogen	Pb	Lead
Ca	Calcium	Hg	Mercury	Rn	Radon
Cd	Cadmium	K	Potassium	S	Sulphur
Cl	Chlorine	Mg	Magnesium	Se	Selenium
Cr	Chromium	Mn	Manganese	Si	Silicon
Cs	Caesium	N	Nitrogen	U	Uranium
Cu	Copper	Na	Sodium	Zn	Zinc
F	Fluorine	O	Oxygen		

Symbol	Name
$CaCO_3$	Calcium carbonate, limestone
$CaSO_4$	Calcium sulphate
$C_6H_{12}O_6$	One of the monosacharide sugars, glucose or fructose
CH_4	Methane
CO	Carbon monoxide
CO_2	Carbon dioxide
HCl	Hydrochloric acid
H_2O	Water
H_2S	Hydrogen sulphide
H_2SO_4	Sulphuric acid
NO_x	Nitrogen oxides, includes NO and NO_2
N_2O	Nitrous oxide
NO	Nitric oxide
NO_2	Nitrogen dioxide
NO^-_2	Nitrite
NO^-_3	Nitrate
NH_3	Ammonia
NH_4	Ammonium
O_3	Ozone
SO_2	Sulphur dioxide
SO_3	Sulphur trioxide

The knack is to read chemical formulae as sentences. The following sentence is straightforwardly factual: 'Erosion of limestone used as building stone is caused by acidified rainfall, in the form of sulphuric acid, reacting with the limestone to form the weaker, more erodible calcium sulphate, carbon dioxide and water'.

This next line says the same thing in more scientific notation. The trick is to substitute chemical symbols for words:

$$H_2SO_4 + CaCO_3 \rightarrow CaSO_4 + CO_2 + H_2O$$

(Sulphuric acid + limestone reacts to give Calcium sulphate + Carbon dioxide + water)

Easy! Chemical symbols are like acronyms, a substitute for words. Now it's your turn. Practise with **Try This 28.2**.

Read the following formulae as sentences. What do these relationships describe and what are their GEES meanings? Answers on p 306.

$CH_4 + O_2 \rightarrow CO_2 + 2H_2O$

$NH_3 + H_2O \div NH_4 + OH$

$SO_3 + H_2O \rightarrow H_2SO_4$

$12H_2O + 6CO_2 + 709 \text{ kcal} \rightarrow C_6H_{12}O_6 + 6O_2 + 6H_2O$

In chemical equations subscripts represent the number of atoms of each element of the molecule. The coefficients (large number) represent the number of molecules of the substance in the reaction. Balance equations by changing the coefficients. Never change subscripts.

28.13 Dyslexia and dyscalculia

Approximately 16 per cent of students at the university are disabled in some way. The largest group are dyslexic or have dyscalculia. Students who had support for dyslexia and dyscalculia through school bring their assessment report to university, and the support continues. If you are aware of difficulties get assessed at university.

Most lecturers understand that most dyslexic students work harder than their colleagues, because they realize that they need to be more organized, take more time over writing, reviewing and reading, and take time to think about what they want to say. It is normal for the students to ask other people to help with proof-reading an essay, and to be good with spell checkers and grammar checkers. One definition of graduateness is taking responsibility for your own learning which is usually very evident in the way dyslexia students organize themselves.

Check out your own university site, or any of these – all accessed 15 February 2011.

Dyslexia at College http://www.dyslexia-college.com/

Dyslexia Help http://www.dyslexiahelp.co.uk/

Hargreaves S 2007 *Study Skills for Dyslexic Students*, Sage, London

Manchester Metropolitan University Dyslexia http://www.mmu.ac.uk/academic/studserv/learningsupport/dyslexia/

Newcastle University http://www.ncl.ac.uk/students/wellbeing/disability-support/dyslexia/

Remember that the vast majority of students who have learning support needs graduate and go on to get graduate jobs.

28.14 An acronym starter list

GEES students compile their personal acronym list, depending on their interests. This is a start; add acronyms to it to make your own resource.

AI	Artificial Intelligence (unless it is an agriculture lecture!)
AOD	Above Ordnance Datum
AONB	Area of Outstanding Natural Beauty
ASCII	American Standard Code for Information Exchange
BATNEEC	Best Available Techniques Not Entailing Excessive Cost
BOD	Biological Oxygen Demand
BPEO	Best Practicable Environmental Option
BTCV	British Trust for Conservation Volunteers
CARICOM	Caribbean Community
CBR	Crude Birth Rate
CDR	Crude Death Rate
CD-ROM	Compact Disc Read-Only Memory
CFCs	Chlorofluorocarbons
CIS	Commonwealth of Independent States
COD	Chemical Oxygen Demand
CSO	Combined Sewer Overflow
DEM	Digital Elevation Model
DO	Dissolved Oxygen
DSS	Decision Support System
DGSM	Digital Geoscientific Spatial Model
DTM	Digital Terrain Model
ECLA	Economic Commission for Latin America
EIA	Environmental Impact Assessment
ESRI	Environmental Systems Research Institute
FAO	Food and Agriculture Organization, part of United Nations Organization
FAQ	Frequently Asked Questions
FTP	File Transfer Protocol
FWAG	Farming and Wildlife Advisory Group
GATT	General Agreement on Tariffs and Trade
GNP	Gross National Product
HCFCs	Hydrochlorofluorocarbons
HSE	Health and Safety Executive
HTML	Hyper Text Markup Language

IMF	International Monetary Fund
IPC	Integrated Pollution Control
LULU	Locally Unacceptable Land Use (nuclear waste, AIDS hospice, see NIMBY)
MBC	Metropolitan Borough Council
MAFF	Ministry of Agriculture Fisheries and Food
MCQs	Multiple Choice Questions
NATO	North Atlantic Treaty Organization
NGOs	Non-Governmental Organizations
NIC	Newly Industrialized Country
NIMBY	Not In My Back Yard
NIR	Natural Increase Rate
NPP	Net Primary Production
NSA	Nitrate Sensitive Area
NVZ	Nitrate Vulnerable Zone
OAS	Organization of American States
OAU	Organization of African Unity
OECD	Organization for Economic Co-operation and Development
OPEC	Organization of Petroleum Exporting Countries
PCBs	Polychlorinated biphenyls
PCV	Prescribed Concentration Value
pH	Logarithmic scale measuring acidity and alkalinity
ppb	parts per billion
R&D	Research and Development
RCS	Rock Classification Scheme
SPSS	Statistical Package for the Social Sciences
SSSI	Site of Special Scientific Interest
TDS	Total Dissolved Solids
TFR	Total Fertility Rate
TLV	Threshold Limiting Value
TNC	Transnational Corporation
TOC	Total Organic Carbon
UNESCO	United Nations Educational, Scientific and Cultural Organization
UNHCR	United Nations Commission on Human Rights
VOCs	Volatile Organic Compounds
WYSIWYG	What You See Is What You Get
WHO	World Health Organization, part of United Nations
WMO	World Meteorological Organization

28.15 Mature students (not to be read by anyone under 20)

They are all so young, and so bright and I don't think I can do this.

Oh yes you can. You have had the bottle to get your act together and make massive arrangements for family and work, handling a class of bright-faced 19-year-olds is easy. Keep remembering that all those smart teenagers have developed loads of bad study habits at school, are out partying most nights, are fantasizing about the bloke or girl they met last night/want to meet and haven't got your incentives to succeed. Your degree is taking time from other activities, reducing the family income and pension contributions; great motivations for success. Most mature students work harder than students straight from school and do very well.

'I really didn't want to say anything in tutorial, I thought they would laugh when I got it wrong'. I'm nervous, you are nervous, he ... (conjugate to gain confidence or get to sleep). At the start of a course, everyone in the group is nervous. Your experience of talking with people at work, home, office, family, scout group ... means YOU CAN DO THIS. It is likely that your age gets you unexpected kudos; younger students equate age with experience and are likely to listen to and value your input.

'I hated the first week, all those 18-year-olds partying and I was trying to register and pick up the kids'. The first term is stressful, but having made compromises to get to university, give it a go, at least until the first set of examinations. If life is really dire, you could switch to part-time for the first year, gaining time to get used to study and the complexities of coping with home and friends.

Mature students end up leading group work more often than the average. Your fellow students will be very, very, very, very happy to let you lead each time as it means they can work less hard. Make sure they support you too!

'It takes me ages to learn things, my brain is really slow'. OK, recognize that it does take longer for older brains to absorb new ideas and concepts. The trick is to be organized and be ACTIVE in studying. This whole book should prompt useful ideas. The following suggestions are gleaned from a selection of mature students. It is a case of finding the tips and routines that work for you. Like playing the cello, keep practising.

Keep fit!

TOP TIPS

☺ Review notes the day after a lecture, and at the weekend.

☺ Check your notes against the texts or papers; ask 'do I understand this point?'

☺ Practise writing regularly, anything from a summary paragraph to an essay. Write short paragraphs, which summarize the main points from a lecture or reading, for use in revision. Devising quizzes is fun (p 223).

☺ Talk to people about your study topics – your friends, partner, children, people you meet at bus stops, the hamster. It gives great practice in summarizing material and explaining in an interesting manner raises your own interest levels. People often ask useful questions.

☺ Visit the library, physically or online, on a regular basis: timetable part of a day or evening each week.

☺ Meet a fellow mature student once week or fortnight to chat about experiences and coursework over lunch, coffee or a drink.

☺ Have a study timetable, and a regular place to study – the shed, attic, bedroom or a corner of the hall. Once the family knows you are out of sight, they will find a devious things to do unsupervised. Stick a notice on the door: KEEP OUT, MUM'S WORKING!

☺ Make specific family fun time and stick to it.

☺ Remind yourself why you decided to do this degree, then ask: is this statistics or computer practical really worse than anything else I have ever had to do? OK, so it is worse, but it will be over in X weeks.

Some time management and a bit of organization will get you through and by the second year you will know so much more about the subjects and about how to manage. When you have finished and got your degree, your family will be amazed, stunned and you will have the qualifications you want. Don't feel guilty; lifelong learning is the way the world is going, and there is increasing provision and awareness of mature students' needs.

> As a family we really tried to go out once a week for an hour, an evening, or all day. This made a real difference. We did walks, swimming, supper at the pub, visits to friends and family and loads of odd things. We took turns to choose where we went, which the kids thought was great.

Problems?

Check out the Students' Union. Most union welfare offices have advice sheets for mature students, advisors for mature students and a mature students society. If there is no mature students society, start one. Generally, GEES staff are very experienced in student problems and solving them. Most departments have a staff tutor or contact for mature students. Go and bother this nice person sooner than later. If you feel really guilty, buy him or her a drink sometime!

Families

Even the most helpful family members have other things in their heads besides remembering that they promised to clean out the rabbit, vacuum the bedroom or buy lemons on the way home. On balance, a list of chores that everyone agrees to and adheres to on 40 per cent of occasions, is good going. Thank the kids regularly for getting jobs done, and remember this degree should not take over all their lives too. YOU NEED to take time away from study too, so plan in trips to the cinema, games of badminton, visits to sports centres and get away from academic activities.

> *I was pretty nervous, but the lecturers were friendly and pleased to see you.*

Short courses

If things get really rough, it is possible to suspend studying for a semester or a year, or stop after one or two years. Most universities offer some certification for completion of each whole or part year. See your university and student union website, and:

Open University 2011 Mature students and CVs: FAQs, http://www.open.ac.uk/careers/mature-students-and-cvs-faqs.php   Accessed 15 February 2011

University of Sheffield 2011 Essential Guide for Mature Students, http://www.sheffield.ac.uk/ssid/welfare/mature/index.html   Accessed 15 February 2011

28.16 Dropping out

'I've been here six weeks, no one has spoken to me and I hate it, I'm off'. Happily, this is a rare experience for GEES students, but every year there are a few people among the thousands taking GEES degrees who are not happy. The main reasons are 'wrong subject choice, I should have done ...', 'Everyone else is cleverer than me', ' the course was not what I expected', 'I was so shattered after A level and school, I really need a break and a rest' and 'I really felt I didn't fit in and it wasn't right'. Homesickness is usual. All GEES departments lose students for these kinds of reasons; you will not be the first or last student feeling unsettled. There are tutors and people in the students union to advise you. Talk to someone as soon as you start to feel unhappy, waiting will probably make you feel worse.

The good news is that most people survive the first weeks, get involved with work, social and sporting activities, and really enjoy themselves. Remember, there are 6000–50,000+ people on campus, 99.9 per cent are very nice and at least 99.7 per cent feel as shy as you do. If you are really at odds with university life, explore the options of suspending your studies, taking a year out or transferring to another university. Take the time to consider options, don't rush your decision.

Question all unsupported statistics!

28.17 Equal opportunities, harassment, gender and racial issues

All universities have equal opportunities policies; they aim to treat students and staff fairly and justly. Happily, most of the time there are few problems. This section cannot discuss these issues in detail, but if you feel there is a problem, please seek advice. Talk to someone sooner rather than later; there will be a tutor who oversees these issues. Your problem is unlikely to be a new one. There is a great deal of advice and experience available, tap into it. The student union, university or college handbook will have contact names for the university equal opportunities officer, or adviser on equal opportunities.

See your own university websites, or visit:

University of Oxford 2011 Equal Opportunities Policy for Students, http://www. admin.ox.ac.uk/eop/statements.shtml   Accessed 15 February 2011

28.18 Stress

All students suffer from stress; it is normal, but needs management. Stress is a bodily reaction to the demands of daily life, and arises when you feel that life's physical, emotional or psychological demands are getting too much. Some people view stress as a challenge and it is vital to them for getting jobs done; stressful events are healthy challenges. Where events are seen as threatening, then distress and unhappiness may follow. If you are feeling upset, or find yourself buttering the kettle and not the toast, try to analyse why. Watch out for circumstances where you are stressed because of:

- Expectations you have for yourself (*Are they reasonable at this time in these circumstances?*)
- Expectations of others, especially parents and tutors (*They have the best motives, but are these reasonable expectations at this time?*)

- Physical environment – noisy flatmates, people who don't wash up, wet weather, hot weather, dark evenings (*What can be done to ameliorate these stresses?*)
- Academic pressures, too many deadlines, not enough time to read (*Can you use study buddies to share study problems? Would it help to talk to someone about time management?*)
- Social pressures, partying all night (*Are friends making unreasonable demands?*)

Serious stress needs proper professional attention, no text will substitute. How do you recognize stress in yourself or friends? Watch for signs like feeling tense, irritable, fatigued, depressed, lacking interest in your studies, having a reduced ability to concentrate, apathy and a tendency to get too stuck into stimulants like drink, drugs and nicotine. So that covers most of us; DO NOT GET PARANOID. PLEASE do not wander into the health centre waving this page, and demand attention for what is really a hangover following a work-free term. Thank you.

Managing stress effectively is mostly about balancing demands and desires, getting a mixture of academic and jolly activities, taking time off if you tend to the 'workaholic' approach (the workaholic student spends nine hours a day in the library outside lectures, has read more papers than the lecturer and is still panicking). Get some exercise – aerobics, line dancing, jogging, swimming, any sport – walk somewhere each day, practise relaxation skills – Tai Chi classes are great fun. Study regularly and for sensible time periods. Break down big tasks into little chunks and tick them off as you do them. If you are still stressed, TALK TO SOMEONE and visit your university health centre.

See your university website or take a look at:

Oxford Brookes University 2011 Managing stress, http://www.brookes.ac.uk/ student/services/health/stress.html   Accessed 15 February 2011

28.19 We hope you don't need: UK helplines

Most people have a great time at university, but if you or your friends have problems, here are some starting points. Non-UK students should search for equivalent organizations using key words, although the information from these sites may prove helpful. (All accessed 25 February 2011.)

Alcohol Concern http://www.alcoholconcern.org.uk

Depression Alliance http://www.depressionalliance.org/

Drugscope http://www.drugscope.org.uk

Release, drugs, the law and human rights http://www.release.org.uk

Samaritans http://www.samaritans.org/

Seasonal Affective Disorder http://www.sada.org.uk/

Social Anxiety http://www.social-anxiety.org.uk/

The Terrence Higgins Trust, HIV and AIDS Charity http://www.tht.org.uk/

Women's Aid http://www.womensaid.org.uk/

CARPE DIEM

or

'May the Force be with you'

Yoda

Try This, crossword and quiz answers

Geojumble 1 p 14

Scotland, Granite, Cyclone, Oxygen

Try This 2.4 – Using PDP reflection in modules p 21

A selection of answers from Level 2 students:

What I want to get out of attending this module is ... *I want to realize my full communication and research skills/I would like the module to give me a clearer insight into a topic I enjoy and think I might want to pursue as a career/I want to improve my computing skills.*

I have discovered the following about myself with respect to decision-making ... *I now realize that I make more decisions than I realize, but in general I try to avoid the process if possible/I am not particularly decisive, but when I have to make a decision I think things through very carefully/I adopt different decision making processes in different cases/I know it is a very weak point/I think I will find it helpful to understand more about how I make decisions in order to improve. I probably agonise too much/Sometimes I am really rational and think things through, other times I am totally impulsive.*

The skills the group used well were ... *As a group we had good discussions, no one person took on the role of leader. Everyone listened to everyone else and everyone had valid points to make/I took part in the discussion, however some members of the group did this better.*

Our group made decisions by ... *general agreement, but mostly one person took the lead and we all just went along with it. We had some texts that got us confused and no one was doing much until the last week. We didn't look at ...(VLE page) which we realized at the end would have helped. Looking back my three words to describe us are chaotic, muddled, unfocused – and we got 64% – result!*

The preparation for the group presentation was ... *To start with we had a distinct lack of preparation, hadn't read the briefing material, and generally didn't know where to start. We overestimated the degree of detail required which explains why we took so long. Organizing what we were supposed to do and deciding who should do what, wasted some time.*

What did you enjoy least about the exercise/session/module/degree course? *There was a lot of information to evaluate in a short time, a bit of quick thinking required/ some aspects were boring and appeared unnecessary/managing time was difficult.*

Quick crossword 1, p 26

Across: 1 Spruce, 5 Time, 8 Lamb, 9 Tailings, 10 Reindeer, 11 Peat, 12 Humber, 14 Stream, 16 Scan, 18 Northern, 20 Upthrust, 21 Byte, 22 Ness, 23 Saturn.
Down: 2 Plateau, 3 Urban, 4 Entrepreneur, 5 Tripper, 6 Magma, 7 Cirrostratus, 13 Benthos, 15 Aerator, 17 Copse, 19 Habit.

Words in Geo-words 1 p 34

There are over 100 words in SEDIMENT. Thirty is a good score, some words can appear twice in the list with and without a final *s*. You may have found:

deem, deism, deme, dement, demise, den, denies, denim, dense, dent, diet, din, dine, dies, diet, dint, edit, emend, ends, inset, item, meet, men, mend, mete, mid, midst, mind, mindset, mine, mint, mist, mite, nest, nets, seed, seem, semen, sent, semi, side, sine, site, sited, smite, snide, stem, teem, teen, tee, ten, tend, tense, tide, tied, time, timed, tines, tin.

Geolinks 1, p 44

B	A	L	I		U	I	S	T		B	U	T	E		E	I	R	E
B	A	L	K		L	I	S	T		B	A	T	E		M	I	R	E
B	A	C	K		L	I	N	T		F	A	T	E		M	I	L	E
B	U	C	K		L	I	N	G		F	A	R	E		M	I	L	L
M	U	C	K		L	O	N	G		F	A	R	N		M	U	L	L

Geo-codeword 1, p 56

1	2	3	4	5	6	7	8	9	10	11	12	13
L	P	E	S	D	Z	U	B	X	I	T	W	J
14	15	16	17	18	19	20	21	22	23	24	25	26
G	O	N	A	Q	C	R	M	F	Y	K	V	H

Drop out 1, p 65

Granite and

E	M	E	R	A	L	D
C	A	D	M	I	U	M
V	E	H	I	C	L	E
F	A	L	L	O	U	T
A	E	R	O	B	I	C
P	O	L	L	U	T	E

Wordsearch 1, p 75

Armenia, Belarus, Belgium, Bulgaria, Croatia, Denmark, Estonia, Finland, France, Germany, Greece, Italy, Latvia, Lithuania, Macedonia, Monaco, Norway, Romania, Moldova, Poland, Serbia, Spain, Sweden, Turkey, Ukraine.

Geograms 2, p 86

Savanna, tortile, graphite.
Add up 1, p 90 118, 103

Try This 10.2 – Reasoned statements, p 96

Some responses; can you further improve each one with references and more detail?

1. *The questionnaire results are right.* We can be confident of the inferences drawn from the questionnaire survey, because the sample size was large (N=678), the criteria established to ensure representative sampling were achieved, and analysis showed statistically significant relationships which confirmed the hypotheses.

2. *Pelagic sediments are found in the deep oceans.* Sediments that accumulate on the ocean floor, well away from coast lines are dominated by clay particles and the detritus of marine organisms. These pelagic sediments are generally fine grained with clay, silica and organic oozes as the majority constituents.

3. *Man is inflicting potentially catastrophic damage on the atmosphere and causing worldwide climate change.* Information about climate change is currently speculative, but some scientific results support the hypothesis that man is damaging the atmosphere. Smoggit (2020) suggests that increasing concentrations of CO_2 and infrared absorbing gases released into the atmosphere as waste products of combustion and other human activities, are causing a net rise in global temperatures. Thermals (2020) suggests that one consequence will be higher global temperatures in the next 50 years than previously recorded.

4. *The United States has become an urban country.* At the start of the eighteenth century, 95 per cent of the population of the United States lived in rural communities or on isolated farms (Demoggers 2020). This number declined across two centuries to about 25 per cent, although of the people who live in rural areas as few as 2 per cent are employed in agriculture. The pull of industrialization and the push of mechanization in agriculture were major factors in this switch to an urban culture.

5. *Pedestrianization civilizes cities.* Pedestrianzation can create a safer, less polluted and apparently more spacious city. Freeing the centre of cars reduces CO and noise pollution and makes centres more attractive for pedestrians. Shoppers appreciate the calmer atmosphere, and retail therapy (shopping) is a pleasure rather than a chore (Shoppersareus 2020).

Geo-codeword 3, p 102

1	2	3	4	5	6	7	8	9	10	11	12	13
G	P	U	E	N	R	C	Y	I	W	T	X	D
14	**15**	**16**	**17**	**18**	**19**	**20**	**21**	**22**	**23**	**24**	**25**	**26**
K	O	B	Z	H	M	S	A	Q	F	J	V	L

Try This 11.1 – Logical arguments, p 109

Some responses:

1. Divergent plate margins occur where two plates are moving away from each other, causing sea-floor spreading. [The first part is a reasonable definition of divergent plate margins. The second part is true where the plate margin is under the ocean. Are there divergent plate margins on land? There are examples on Iceland, and what about rift valleys? There is nothing about time scales or the rates involved.]

2. Rainforests store more carbon in their plant tissue than any other vegetation type. Burning forests release this stored carbon into the atmosphere as CO_2. The net result is increased CO_2 in the atmosphere. [Sounds OK, provided the first sentence is factually correct. If it is the density of the vegetation in tropical rainforests which maximizes the carbon content, then the first sentence is untrue. There are other factors to consider too, like harvesting.]

3. If the system produces a net financial gain, then the management regime is successful and the development economically viable. [This is a rather sweeping statement; no time dimension is mentioned. A net profit might indicate successful management, but it might not be the managers that create the profit. What about maximum profit: could the managers do better? The development may be successful this year, but will this always be the case?]

4. Urban management in the nineteenth century aimed to reduce chaos in the streets from paving, lighting, refuse removal and drainage, and to impose law and order. [The grammar here is awry. Is chaos too strong a word? Was the paving chaotic? Was there any lighting? Did urban managers have wider concerns than sorting out the streets? This is the sort of sentence where you know what the writer is trying to say, but it is slightly off target, general and without precise examples.]

Try This 11.2 – Logical linking phrases p 109

Some examples:

and; however; several attempts have been made; conversely; do not agree; if; subsequently; despite the fact; in addition; this has been broadly related to; such data also; these results are opposed to; in other words; coupled with; making a causal relationship; given the pressures to; for instance; one additional mechanism; none the less; such as; how can this; there is evidence; see; for example; though varied in style; but we are not just finding; at the same time; according to theory; in order to understand; there are several problems with; it is more important to; none of this is to deny; there may be other factors.

Wordsearch 2, p 116

Angola, Algeria, Botswana, Burundi, Cameroon, Chad, Congo, Djibouti, Egypt, Eritrea, Ethiopia, Gabon, Gambia, Ghana, Kenya, Lesotho, Libya, Malawi, Morocco, Mozambique, Namibia, Nigeria, Senegal, Somalia, Sudan, Togo, Tunisia, Uganda, Zambia, Zimbabwe.

Add up 2, p 131

127, 131

Try This 14.1 – Working with a brainstormed list, p 134

There are many ways to order this information; the headings here are suggestions. Notice that there is some repetition, a mix of facts and opinions, and some points could go under more than one heading.

- **Slope instability features**: Movement of handrails. Tennis courts on old levelled slip. Evidence that paths across the slips are regularly re-laid, cracked tarmac, fences reset. Promenade held up by an assortment of stilts, walls and cantilever structures where slope below has been undercut. Castle fortification, some walls now undermined and eroded. Path covered by slumping clay. Vegetation eroded from wetter clay slopes.
- **Geology and geomorphology**: Complex geology, land slips, big fault structures. Undercutting of cliffs by by waves. Rocks at cliff foot dissipate wave energy. Soil overlying clay.
- **Slope repair and restoration features**: Concreting the slopes to prevent soil erosion. Fine and coarse netting, geotextiles, used on clay slope to help stabilize the soil. Bolts into the rock face to increase stability. Rock armour installed at the foot of recent slip. Wall to prevent boulders from cliff hitting the road. Tree planting on recent and older slip sites. Drainage holes in retaining walls to reduce soil water pressure.
- **Tourism factors**: Café for tourists and visitors at foot of slip area. Hotel, amusement arcade and tourist shops – noisy and unsightly. Harbour has mix of marina, fishing and tourist fairground rides. Victorian swimming pool, in need of repair.
- **Transport issues**: Car parking demands, unsightly in long views. Car pollution. Car parking unscenic. Caravan park on cliff.

- **Beach issues:** Sea wall prevents undercutting. Rocks at cliff foot dissipate wave energy. Groynes prevent longshore drift. Repairs to sea wall.
- **Other issues**: Tourist facilities destroy foreshore view. Soil overlying clay rock. Vegetation worn on paths and alongside paths. Litter needs collecting from beach. Road built over old stream, presumed piped underground. Varied quality of coastal path, not always well signposted.

Try This 14.3 – Discussant's role, p 139

Positive points: Offers factual information ; Gives factual information; Asks for examples; Encourages others to speak ; Asks for reactions; Asks for examples; Offers opinions; Helps to summarize the discussion; Asks for opinions; Gives examples; Summarizes and moves discussion to next point.

Negative attitudes: Seeks the sympathy vote; Is very competitive; Speaks aggressively; Is very defensive; Keeps quiet; Ignores a member's contribution; Mucks about; Diverts the discussion to other topics; Keeps arguing for the same idea although the discussion has moved on; Is very (aggressively) confrontational.

Quick crossword 2, p 143

Across: 1 Tariffs, 5 Datable, 9 Fecundity, 10 Coypu, 11 Climatologist, 13 Icefield, 15 Linear, 17 Talkie, 19 Diaspora, 22 Rubber stamped, 25 Aorta, 26 Acidifier, 27 Diluted, 28 Mayweed.

Down: 1 Tufa, 2 Recycle, 3 Fungi, 4 Spiracle, 5 Dry rot, 6 Tectonics, 7 Bayside, 8 Equatorial, 12 Hinterland, 14 Itinerant, 16 Titanium, 18 Liberal, 20 Old Time, 21 Island, 23 Privy, 24 Grid.

Words in Geo-words 2 p 156

Here are just a few words found in PALAEONTOLOGY to get you started:
Aeon, agape, agate, aglet, agony, alga, algal, allot, alloy, ally, aloe, alone, analog, angel, apollo, alto, élan, galena, galleon, gallop, goat, gooey, lagoon, lane, lapel, lately, lean, leapt, legal, lento, lonely, loopy, loyal, natal, neap, neatly, oaten, ontology, openly, plant, planet, plate, tango, toga, topology, yang, yoga.

Try This 17.1 – Keywords in essay questions, p 171

1. *Discuss the role and responsibilities of science and scientists for society*. This essay requires a balanced four-part answer addressing roles and responsibilities for science in general and individual scientists. Ethics, knowledge, clarity of explanation, empiricism and its limitations, oversight by communities, funders and governments are just some areas that could be included, with examples and references for each. Remember to use geography, earth or environment

examples as fits your degree. This essay could be set for any science student, Chemistry students will answer with chemistry examples, psychologists with ...

2. *Communications, internal commerce and energy are the sectors that are usually identified as the most serious 'bottlenecks' in contemporary Chinese development. Explain the weaknesses in one of these sectors.* Asks for 'weaknesses' and 'one' sector only. Don't attempt to impress by covering all <u>three</u> sectors <u>and</u> strengths <u>and</u> weaknesses, giving an examiner pages of irrelevant writing to cross out. What is 'contemporary' in this context? Starting with nineteenth-century material will not help.

3. *To what extent are growth and change impeded by archaic social structures in either Latin America or a selected country in Latin America?* Keywords here are 'growth', 'change' and 'impeded', so there should be little or no explanation of how social structures assist growth and change. Take examples from <u>either</u> across Latin America <u>or</u> just one country.

4. *Critically explain how Andrews (1983) derived his bedload entrainment function.* Wants a technical explanation leading to the derivation of an equation in 1983. Therefore, call on pre-1983 work and the evidence Andrews used, and alternatives and arguments against this particular function. Any reference to post-1983 developments is beyond the scope of this essay. You might refer to subsequent developments in a closing paragraph, but very briefly.

5. *Explore the arguments that further drilling for oil should be permitted in Alaska.* This is a very open question, it asks for arguments for drilling, but do not ignore the arguments against drilling. Will you consider all of Alaska, or focus on drilling in the Alaska National Wildlife Refuge, which is topical? The perspectives of many pressure groups are important, do they have valid and evidenced positions? What are the alternatives to drilling in Alaska? This question needs you to decide on the balance of geology, resource availability, economics, politics at national and local level, ethics, landscape and wildlife impacts ...

6. *Evaluate the role of mass media reportage of environmental issues.* In 'evaluating' be complimentary as well as critical, remember the advantages as in raising awareness, as well as the disadvantages. Consider all types of mass media, different types of TV reportage – the news bulletin versus the considered documentary, the print media, papers and magazines. The answer requires diverse examples. Is this a UK question or should comment be made on media activities in other countries? Incidentally, if all the lecture examples were from the UK you can use European and other examples, unless the module is country specific: 'Environmental Issues in the UK'.

Try This 17.2 – Evaluate an introduction, p 173

Some comments – do you agree? I think versions 2 and 4 are of good enough standard for a university essay. Versions 1 and 3 contain correct statements, but resemble a bundle of random thoughts rather than a developed argument. There is much to discuss, accordingly guidelines help the reader. Version 1 starts promisingly, but the last sentence is unconnected and there is nothing to indicate where the essay is going. Version 3 could introduce a wetlands essay in any country – did you notice Canada was not mentioned? Personally, I do not like the definition of wetland in brackets in the first sentence of version 3. It is not wrong exactly, but to my mind, important enough not to be in parenthesis. The title of the essay is effectively repeated in the third and last sentence. Version 2 is much longer and covers considerable ground. The last sentence indicates that evidence is to be marshalled through case evidence and that there is a positive side to man's impact through wetland reclamation and restoration. Version 4 is slimmer: two main points are made in the first two sentences, and then the essay structure is flagged in the last two. Versions 2 and 4 benefit from the embedded references.

Try This 17.3 – Good concluding paragraphs, p 177

Version 1   The tone here is upbeat to the point of tabloidese: 'mega' is not a good adjective and 'explosion' is OTT. The second sentence makes no sense in relation to the first, and ascribes some anthropomorphic attributes to non-sentient wetlands. The fourth sentence does not follow from the third, although using the phrase, 'This has meant', implies a logical link.

Version 2   A single-sentence paragraph is usually a bad idea. The sentiments are right but the language is very relaxed!

Version 3   Best of these!

Version 4   This version starts by sounding more like an introduction and then tails off. Individually the statements are all true, but they don't coalesce.

Version 5   Starting a final paragraph with 'To conclude' wastes words. The reader can see it is the last paragraph. The 'This essay has shown' formulae lacks initiative. Not wrong, but not innovative either. The first sentence makes a geographical generalization. Did man actually have a national policy here, or is the national picture the accumulation of a great many local, independent wetland drainage activities?

Try This 17.6 – Shorten these sentences, p 180

Wordy	Improved
In many cases, the tourists were overcharged.	Many tourists were overcharged.
Microbes are an important factor in soil processes.	Microbes control decay rates in soils.
It is rarely the case that sampling is too detailed.	Sampling is rarely too detailed.
The headman was the proud possessor of much of the land in the vicinity of the village.	The headman owned land around the village OR The headman was proud to own land near the village.
Moving to another phase of the project ...	The next phase ...
Chi-square is a type of statistical test. (should read 'Chi-square is a type of statistical test')	Chi-square is a statistical test.
One of the best ways of tackling prison reform is...	To tackle prison reform ...
The investigation of cross-bedding at Ramsgate continues along the lines outlined.	Investigations of cross-bedding at Ramsgate are continuing.
The nature of the problem ...	The problem ...
Temperature is increasingly important in influencing the rate of snowmelt.	Temperature controls snowmelt rates.
One prominent feature of the landscape was the narrow valleys.	Narrow valleys are prominent landscape features.
It is sort of understood that ...	It is understood that ...
It is difficult enough to learn about stratigraphy without time constraints adding to the pressure.	Learning about stratigraphy is time consuming.
The body of evidence is in favour of ...	The evidence supports ...

Try This 17.8 – Synonyms for GEESers, p 182

1 understanding, 2 main, 3 discrimination, 4 extracted, 5 explains, 6 estimates, 7 experience, 8 associated, 9 Could be either, choose morphology as the more technical term.

Geojumble 3, p 186

Solvent, Harvest, Emission

Add up 3, p 199

179, 106

Drop Out 2 p 196

Mexican, and mammoth, Permian, gumtree, plateau, isotope.

Words in Geo-words 3, p 202

There are about 80 words to be found in FIELDWORK, with -er and -ed variations. Twenty is a good score, Well done. You may have found some of these: deli, dike, dire, dirk, doer, dowel, dower, drew, fie, field, file, filo, fiord, fire, fled, flew, flier, floe, florid, flow, foil, folder, folk, ford, fried, idle, idol, kilo, led, lido, lied, life, like, lire, lode, lord, lore, low, lower, oiled, older, ole, oriel, owe, owed, red, redo, refold, rid, ride, rife, rifle, rod, rode, roe, role, row, rowed, wed, weir, weird, weld, wide, wield, wife, wired, wok, woke, wolf, word, wore, work, world.

Interlinks 1, p217

Conservationist, northeastward, vulcanology, Triassic, France, Goa.

Try This 22.5 – Good revision questions, p 225

1. What is the purpose of ... ? (cryosphere, trilobite, public–private partnerships)
2. Why is ... an inadequate explanation of ... ?
3. Who are the three main authors to quote for this topic?
4. Name two examples not in the course text or lectures for ..?
5. What are the main characteristics of ... ?
6. Outline the relationship between ... and ...
7. What are the limitations of the ... approach? (core-periphery perspective, nebular hypothesis, ice-cores as climate records)
8. What methodology is employed to ... ?
9. At what point does this process become a hazard?/of concern? (algal blooms, monopolies)
10. How has human impact affected ... ?
11. What was the main aim of ... ? (open-ended questions, reconstructing past climates)
12. What is the spatial scale involved ... ?
13. How is ... calculated? (isotope decay, Reynolds number, market potential)

14. Outline the sequence of events involved in ... ?
15. What are the implications of ... not occurring? (fire in ecosystems, inward investment)
16. How has ... adapted to ... ? (vegetation in arid environments, housing tenure post-war Britain)
17. Outline the different approaches that can be taken to the study of ...
18. Will this work in the same way at a larger/smaller scale?
19. How influential has ... been?
20. What are the ... (regional, national, cultural) ... implications of ... ?
21. What is meant by ..? (commensalism, regional metamorphism, logical positivism, crude death rate, safety factor)
22. Why is ... important for ... ? (copper ... plant growth, secondary census data ... population studies)

Wordsearch 3, p226

Cyclone, danger, drought, earthquake, epidemic, eruption, explosion, exposure, flood, hazard, monsoon, pandemic, plague, pollutant, risk, storm, threat, tornado, tsunami, typhoon.

Wordsearch 4, p 226

Alluvium, basalt, carbon cycle, carbonate, cloud, endangered species, energy, fjord, fumerole, gabbro hematite, hydrocarbon, karst, lithosphere, metamorphism, ocean, parasite, pesticide, radiation, recycled, salinity, wildlife.

Interlinks 2, p 241

Humanitarianism, geomorphology, sheepshearer, tectonic, cosmos, bay.

Words in Geo-words 4, p249

BASALT: Alba, atlas, Baal, balsa, basal, blast, blat, lab, last, salt, slab, stab, tabs.

Drop Out 3, p 258

Mercury and Sulphur, Ethanol, Density, Capital, Phenols, Entropy.

Words in Geo-words 5, p 271

TECTONIC. Fifteen is a good score. You may have found: cent, cine, cite, coin, conceit, cone, conic, cot, cote, eon, ice, icon, into, net, nett, nice, not, note, notice, octet, one, once, ten, tent, tin, tinct, tine, tint, ton, tone, tonic, tot, tote.

Try This 28.2 – Chemical formulae, p 286

$CH_4 + O_2 \rightarrow CO_2 + 2H_2O$ (When methane and oxygen are burned together the reaction gives carbon dioxide and water, a greenhouse gas issue.)

$NH_3 + H_2O \rightarrow NH_4 + OH$ (Describes the ammonia to ammonium oxidation – reduction interaction, a soils process.)

$SO_3 + H_2O \rightarrow H_2SO_4$ (An atmosphere interaction where sulphur trioxide and water combine to form sulphuric acid, which causes acid rain.)

$12H_2O + 6CO_2 + 709\ kcal \rightarrow C_6H_{12}O_6 + 6O_2 + 6H_2O$ (Describes the synthesis of water and carbon dioxide with light energy in plant leaves to produce the carbohydrate $C_6H_{12}O_6$ for plant growth with water and oxygen as waste products – a biogeography issue.)

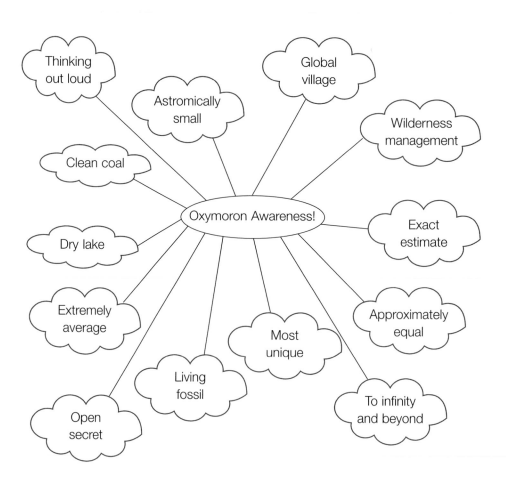

Index